Don,

Thank you for participating in FACM08 Conference to celebrate my 75th birthday. It means a lot to me and my family.

Best Regards,

Dafo

Frontiers of APPLIED AND COMPUTATIONAL MATHEMATICS

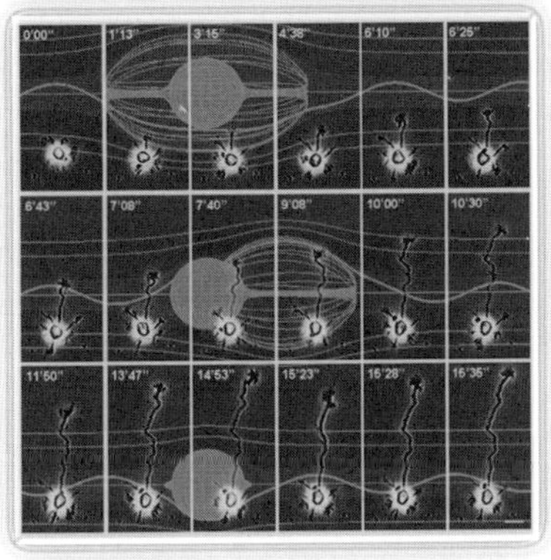

Frontiers of APPLIED AND COMPUTATIONAL MATHEMATICS

New Jersey Institute of Technology, USA 19–21 May 2008

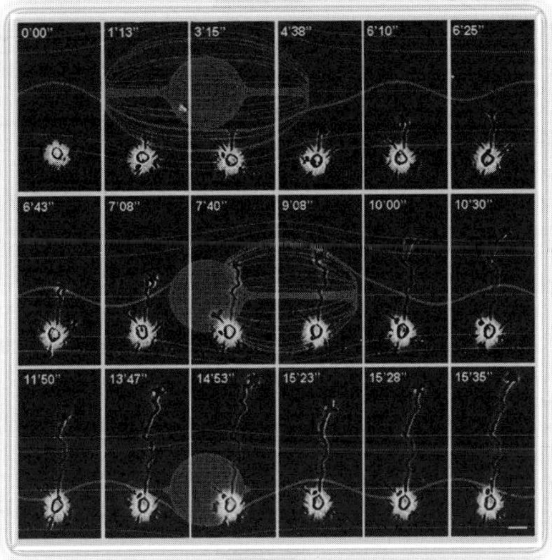

edited by

Denis Blackmore • Amitabha Bose • Peter Petropoulos
New Jersey Institute of Technology, USA

World Scientific

NEW JERSEY • LONDON • SINGAPORE • BEIJING • SHANGHAI • HONG KONG • TAIPEI • CHENNAI

Published by

World Scientific Publishing Co. Pte. Ltd.

5 Toh Tuck Link, Singapore 596224

USA office: 27 Warren Street, Suite 401-402, Hackensack, NJ 07601

UK office: 57 Shelton Street, Covent Garden, London WC2H 9HE

Library of Congress Cataloging-in-Publication Data

Conference on Frontiers of Applied and Computational Mathematics (5th : 2008 : New Jersey Institute of Technology)

Frontiers of applied and computational mathematics : dedicated to Daljit Singh Ahluwalia on his 75th birthday : proceedings of the 2008 Conference on FACM-08, New Jersey Institute of Technology, USA, 19–21 May 2008 / edited by Denis Blackmore, Amitabha Bose & Peter Petropoulos.

p. cm.

Includes bibliographical references and index.

ISBN-13: 978-981-283-528-4 (hardcover : alk. paper)

ISBN-10: 981-283-528-8 (hardcover : alk. paper)

1. Applied mathematics--Congresses. 2. Computer science--Mathematics--Congresses. I. Ahluwalia, Daljit Singh, 1932– II. Blackmore, Denis L. III. Bose, Amitabha. IV. Petropoulos, Peter G. V. Title.

T57.N37 2008
601'.51--dc22

2008039284

British Library Cataloguing-in-Publication Data

A catalogue record for this book is available from the British Library.

Printed in Singapore by World Scientific Printers

PREFACE

The fifth annual *Frontiers in Applied and Computational Mathematics Conference* (FACM '08) was held at the New Jersey Institute of Technology in Newark (NJIT), NJ, USA, May 19-21, 2008. This year's conference was one of the largest in the series and was focused on mathematical biology, mathematical fluid dynamics, applied statistics, and wave propagation; areas of research strength at NJIT. The present volume contains the texts of many of the invited talks delivered at the conference.

A special feature of the fifth conference was a celebration of the 75th birthday of Daljit S. Ahluwalia, in honor of his many contributions to the Department and the University. Professor Ahluwalia is longtime Chair of the Department of Mathematical Sciences and Director of the Center for Applied Mathematics and Statistics at NJIT.

The FACM Conferences have been organized as forums for free exchange of ideas and results at the frontiers of research in the mathematical sciences. An overriding aim of is to show the diversity of difficult problems in the biomedical, physical, and social sciences, and in engineering and technology, and how they can be treated using mathematical modeling. In addition to plenary and invited presentations by distinguished scholars, the fifth conference featured 20 contributed presentations by graduate students and postdoctoral researchers.

Thanks are due to President R. Altenkirch, Provost P. Nelson, Senior Vice President D. Sebastian, and Dean F. Deek, and other members of the NJIT administration and staff for their assistance. Financial support for this conference has was provided by NJIT's Strategic Initiative, the Society for Mathematical Biology, and the National Science Foundation.

In addition, the Local Committees members and staff deserve great thanks for creating a well-run and productive conference. It is a pleasure to thank all of them for their outstanding efforts.

Michael Siegel
(Chair, FACM '08 Organizing Committee)

Newark, NJ, U.S.A.
30 May 2008

ORGANIZING COMMITTEES

ORGANIZING COMMITTEE

for Frontiers in Applied and Computational Mathematics 2008

Manish Bhattacharjee	– NJIT, Newark, NJ, USA
Amitabha Bose	– NJIT, Newark, NJ, USA
Gregory A. Kriegsmann	– NJIT, Newark, NJ, USA
Zoi-Heleni Michalopoulou	– NJIT, Newark, NJ, USA
Robert M. Miura	– NJIT, Newark, NJ, USA
Demetrios Papageorgiou	– NJIT, Newark, NJ, USA
Michael Siegel (Chair)	– NJIT, Newark, NJ, USA

CONTRIBUTED PAPERS SCREENING COMMITTEE

Denis Blackmore	– NJIT, Newark, NJ, USA
Pam Cook	– University of Delaware, Newark, DE, USA
Linda Cummings	– University of Nottingham, UK

TRAVEL AWARDS COMMITTEE

Pam Cook	– University of Delaware, Newark, DE, USA
Sunil Dhar	– NJIT, Newark, NJ, USA
Zoi-Heleni Michalopoulou	– NJIT, Newark, NJ, USA

ORGANIZING COMMITTEE STAFF

Sherri Brown, Nathalie Caparo, Fatima Ejallali , Padma Gulati, Susan Sutton – NJIT, Newark, NJ, USA

Daljit Singh Ahluwalia

DEDICATION

FACM'08 was dedicated to Professor Daljit Singh Ahluwalia on his 75th birthday. This was to pay tribute to his extraordinary academic career; in particular, his leadership in guiding the Department of Mathematical Sciences at NJIT to prominence in applied mathematics, as evidenced by the 2008 Academic Analytics ranking in the top 10 among similar departments nationally, based on faculty scholarly productivity.

Daljit S. Ahluwalia was born in Sialkot, India on September 5, 1932. He grew up in Amritsar, the holy city of the Sikhs. Being the oldest male of his family, he was only fifteen when he began his first job. He graduated from high-school in 1948, obtained bachelors and masters degrees in mathematics in 1952 and 1955, respectively, from Punjab University, and went on to teach undergraduate mathematics at Ramgharia College, Phagwara from 1955-1957 and undergraduate and graduate mathematics at Khalsa College and Bombay University from 1957-1962.

He arrived in the United States on a Woodrow Wilson Fellowship to pursue a PhD in applied mathematics at Indiana University, which he completed in three years along with a masters degree in physics in 1965. He then served there for a year as an assistant professor.

In 1966, Professor Ahluwalia accepted a visiting membership at NYU's Courant Institute of Mathematical Sciences for two years. From 1968-1986, he was an Assistant, Associate, and finally a Professor at the Courant Institute. From 1968-1972, he assisted J.B. Keller in establishing the Applied Mathematics Department at the University Heights campus of NYU. After the University Heights Campus closed, he went on to develop a PhD program at the University of South Florida from 1972-1974. He then returned to the Courant Institute where he pursued research in asymptotic methods, wave propagation and acoustics.

He has also been a Visiting Professor at Stanford, Northwestern and Columbia University during his sabbaticals. In 1968, he was invited to participate in the Battelle Memorial Conference in Seattle, Washington, which was also attended by only fifteen mathematicians and an equal number of physicians from all over the world. He also was an active participant in the Office of Naval research sponsored Conferences of the Summer Institute of Applied Mathematics held from 1972-1976, and was a consultant at the Lamont Doherty Geological Observatory, Columbia University.

Professor Ahluwalia's accomplishments outside the academic sphere also have been substantial, especially in service to the Sikh community. In 1971, he was elected General Secretary of the Sikh Cultural Society which built the first Sikh Temple on the East coast. He helped organize the First Sikh Youth camp in the US in 1975, was a founding fellow of the Sikh Renaissance Institute (SRI) in 1976 and the Director of its Educational activities of Sikh youth and its social services for a couple of years, and was elected President of SRI. As Chairman of the Finance Committee of the first India Day Parade in New York in 1981, he directed a very successful fund raising campaign.

As founding President in 1979 of the Overseas Indian Congress of North America, he organized conventions and seminars on "Indo-American Relations", the "Message of Mahatma Gandhi", and other related topics. These were addressed by such distinguished personalities as Indian Ambassador (and later President) K.R. Narayanan; Narasimha Rao, who became Prime Minister of India; Dr. Gerard Piel, President of Scientific America; prominent New York politician, Howard Samuels; Congressman Steven Solarz; Nancy (Mrs. Zubin) Mehta; and Mary Carass, Professor of Political Science at Rutgers University. Subsequently, he was elected a member of the executive committee of Heart and Hand for the Handicapped Children, which raised funds for handicapped children in India and the USA. In recognition of these good works in service to India and America, National Investment and Finance weekly awarded him their Gold Medal and title "Ambassador of Goodwill" in 1983.

He joined NJIT in 1986 as Professor and Chair of the Department of Mathematics after receiving a commitment of support for his vision of building a world-class department, and has served in this position from 1986-89 and 1996-. Under his leadership, the Mathematical Sciences Department has become a nationally recognized center of excellence in research and education, significantly increased its funded research, and has added several successful programs: a B.S. program in Applied Mathematics in 1989, a PhD in Mathematical Sciences in 1995, a M.S. in Applied Statistics in 1997, and both an M.S. in Biostatistics and a B.S. in Computational Sciences in 2008. In addition, he is a founding Director of the research Center in Applied Mathematics and Statistics (CAMS); and has held this position since 1986.

The great strides made by the Department of Mathematical Sciences at NJIT under the stewardship of Professor Ahluwalia, highlighted his leadership abilities, and led to the Department being designated as a Program of Excellence under NJIT's Strategic Initiative. As a result of these and other accomplishments, he was awarded the NJIT Board of Overseers Public and Institute Service Award in 2002, and selected to serve as the Acting Dean of the College of Computing Sciences at NJIT from May, 2005 through June, 2006 after which he returned to his position of Chair of the department of Mathematical Sciences.

Excerpts of some of the tributes Professor Ahluwalia received for his accomplishments are:

> "A special feature of this years conference is a celebration of Daljit S. Ahluwalia's 75th birthday and his contributions to the growth of mathematical sciences at NJIT. An endowed fund for the newly established D. S. Ahluwalia Doctoral Fellowship in Mathematical Sciences will be announced at the conference. Prof. Ahluwalia's major accomplishments as Chair of the Department of Mathematical Sciences include guiding the increase in departmental research funding from nearly nothing to approximately 2 million per year in federal funding, leading the transition from a primarily teaching and service department to a top ten research department as reported in the November, 2007 Chronicle of Higher Education, instituting new degrees in Applied Mathematics, Applied Statistics, Mathematical Sciences at the doctoral level, and, in planning, Biostatistics and Computational Sciences, recruiting outstanding junior faculty who are now leaders in their field and senior faculty with established research records, and establishing a collegial and scholarly atmosphere for study and research that encourages excellence. We all thank Daljit for his many contributions in building one of the most accomplished teams in applied mathematics in the World today."
>
> Robert A. Altenkirch, President
> New Jersey Institute of Technology

"This fifth FACM meeting is excellently organized by the Department of Mathematical Sciences and the Center for Applied Mathematics and Statistics, both in the College of Science and Liberal Arts. The Department is the largest on campus and, in 2004, NJIT selected the Department to receive strategic priority funding to achieve national prominence within a five year period. Under the leadership of Dr. Daljit S. Ahluwalia, the department has made giant strides! The Department is well known nationally and internationally for its excellence, innovation and creativity, and in 2007 was ranked in the top 10 research universities as determined by the faculty scholarly productivity index of the Chronicle of Higher Education."

Priscilla P. Nelson, Provost
New Jersey Institute of Technology

"Professor Ahluwalia's imprints are evident on the entire success story of the Department of Mathematical Sciences (DMS). He has made innumerable contributions not only to his department, but also to our college and the university as a whole since arriving at NJIT in 1986 from the famed Courant Institute of Mathematical Sciences, where he was a professor. Under Professor Ahluwalia's leadership, DMS has transitioned from being a teaching department to one of the leading venues for applied mathematics research in the country. This success has been accomplished by a steady and persistent emphasis on hiring and grooming young faculty, as well as established leading researchers in important applied areas of mathematics: Mathematical Biology, Mathematical Fluid Dynamics, Applied Statistics and Biostatistics, Electromagnetics, and Waves and Acoustics, the very focus areas of FACM'08. Professor Ahluwalia has led DMS by example as any of his colleagues will attest. On a daily basis, he can be found walking the halls of the department, keeping a watchful and nurturing eye on the faculty and students, mentoring his younger faculty, and solving administrative and academic problems in real time."

Fadi P. Deek, Dean
College of Science and Liberal Arts
New Jersey Institute of Technology

"FACM '08 is hosted by the Department of Mathematical Sciences and the Center for Applied Mathematics and Statistics at NJIT. The Department of Mathematical Sciences has been selected under NJIT's Strategic Initiative as one of its programs of excellence that has been designated to attain increased national and international prominence. This fifth FACM conference is partially a result of this Strategic Initiative. A special feature of the fifth conference is a celebration of the 75th birthday of Daljit S. Ahluwalia, in honor of his many contributions to the Department and the University. Daljit is longtime Chair of the Department of Mathematical Sciences and Director of the Center for Applied Mathematics and Statistics. He has played a key role in creating a vibrant and collegial environment in which to pursue research and teaching, and we are fortunate for his vision and legendary energy. Since Daljit arrived in 1986, we've seen the faculty increase by 100%, floors of Cullimore Hall occupied by Mathematical Sciences increase by 300%, and departmental computing power increased 1,000,000 fold."

Michael Siegel, Chair
FACM'08 Organizing Committee

"When Daljit came to the Courant Institute he had not worked on wave propagation, but he quickly learned about it and contributed significantly to our research on problems of propagation and diffraction. He helped to develop a uniform asymptotic theory of edge diffraction, and many other things. Then he turned his attention in a completely different direction, that of developing a first rate department of applied mathematics at NJIT. He succeeded admirably in that endeavor, and the continuing activity of the department will be his enduring legacy."

Joseph B. Keller
Stanford University

CONTENTS

PART A

Plenary Papers

MULTI-SCALE METHODS, COMPUTER SIMULATIONS, AND DATA MINING: DIFFERENCE EQUATIONS AND RENEWAL EQUATIONS

FRANK HOPPENSTEADT*

Courant Institute of Mathematical Sciences, New York University, New York, NY 10012, USA
E-mail: frank.hoppensteadt@nyu.edu

Has applied mathematics disappeared behind the computer screen? High level computer languages and high capacity computers for simulation followed by data mining give powerful tools for studying complex systems. The two examples presented here are a system of difference equations from population genetics and a system of Volterra integral equations from demography, both perturbed by random noise. We analyze these systems using averaging methods and then compare the derived solution with computer simulations of the system. In the first case, mathematical analysis guides mining of the simulated solution data base. In the second, mathematical analysis reveals deterministic chaotic behavior in the averaged system that confounds the simulation/data-mining approach.

Keywords: Averaging; Bogoliuboff Approximations; Monte Carlo Simulations; Volterra Integral Equations.

1. Discrete Event Systems with Noise

We consider two examples where multi-time methods can help to interpret computer simulation of complex systems.

1.1. *Averaging Difference Equations*

Consider a nonlinear discrete event system

$$x_{n+1} = Ax_n + \varepsilon f(n, x_n, \varepsilon),$$

$$x_0 \text{ is given.}$$

*Dedicated to Daljit S. Ahluwhalia on the occasion of his 75th birthday. This work was supported in part by SRC/FCRP/FENA grant 2003-NT-1107.

for $n = 0, 1, 2, 3, \dots$, where $x \in E^N$, $A \in E^{N\times N}$ and is invertible, $f(n, x, \varepsilon)$ is a smooth function of $x \in E^N$ for each value of $n \in \mathcal{N}$, and where ε is a small positive parameter. Since A is invertible, we can define $x_n = A^n z_n$ and get

$$z_{n+1} = z_n + \varepsilon A^{-n-1} f(n, A^n z_n, \varepsilon),$$

so without loss of generality, we replace A by the identity matrix and rewrite f accordingly:

$$x_{n+1} = x_n + \varepsilon f(n, x_n, \varepsilon),$$

$$x_0 \text{ is given.}$$

A two-timing perturbation method suggests a result, which is later proved to be valid. Let $s = \varepsilon n$ and $x_n = u(n, s, \varepsilon)$. We seek the solution in the form

$$u(n, s, \varepsilon) = u_0(n, s) + \varepsilon u_1(n, s) + \dots.$$

Substituting this in the equation gives

$$u(n+1, s+\varepsilon, \varepsilon) = u(n, s, \varepsilon) + \varepsilon f(n, u(n, s, \varepsilon)).$$

Expanding in powers of ε gives

$$\begin{aligned} u_0(n+1, s) &= u_0(n, s) \\ u_1(n+1, s) + \frac{\partial u_0(n, s)}{\partial s} &= u_1(n, s) + f(n, u_0(n, s), 0) \\ &\text{Etc.} \end{aligned}$$

The first equation indicates that u_0 is independent of n; we write

$$u_0(n, s) = U_0(s).$$

The second equation gives

$$u_1(n+1, s) = u_1(n, s) + f(n, U_0(s), 0) - \frac{dU_0}{ds}$$

whose solution is

$$u_1(n, s) = U_1(s) + \sum_{j=0}^{n-1} f(j, U_0(s), 0) - n\frac{dU_0}{ds}$$

where $U_1(s)$ is an arbitrary function.

If u_1 is bounded, then dividing both sides by n and passing to the limit $n = \infty$ gives

$$\frac{dU_0}{ds} = \bar{f}(U_0), \quad U_0(0) = x_0,$$

where

$$\bar{f}(U_0(s)) = \lim_{n\to\infty} \frac{1}{n} \sum_{j=0}^{n-1} f(j, U_0(s), 0).$$

Under natural conditions, we have

$$u_n = U_0(\varepsilon n) + O(\varepsilon).$$

With additional stability conditions on the averaged equation, this approximation can be uniform in $0 \le t < \infty$.[3]

1.2. *Averaging Systems Perturbed by Random Noise*

Next, consider a discrete event system perturbed by random noise:

$$x_{n+1} = x_n + \varepsilon f(x_n, y_n)$$

where y_n is a random process that is stationary and ergodic in a compact space Y, with an ergodic measure m. (Think of simple coin tossing where y takes on one of two values, say 0 and 1, with probability at each trial p_0 and p_1, respectively, so $m(\{0\}) = p_0$ and $m(\{1\}) = p_1$.)

We repeat the above steps

$$x_n = u(n, \varepsilon n, \varepsilon) = u_0(n, s) + \text{ remainder}.$$

As before, we have

$$u_0(n, s) = U_0(s).$$

If $u_1/n = o(1)$ and $f(x, y)$ is a smooth function of x and a measurable function of y, then the ergodic theorem says that the sum above will converge to the average of f over Y: So,

$$\frac{dU_0}{ds}(s) = \lim_{n\to\infty} \frac{1}{n} \sum_{j=0}^{n-1} f(U_0(s), y_j)$$

$$= \int_Y f(U_0, y)\, m(dy) \equiv \bar{f}(U_0).$$

As is usual in such Bogoliuboff approximations, we can show that

$$\frac{x_n - U_0(\varepsilon n)}{\sqrt{\varepsilon}} \approx \text{ diffusion process.}$$

Thus, the approximation to $u(n,\varepsilon)$ involves an average (Law of Large Numbers) and a diffusion process (Central Limit Theorem). Convergence is in a probabilistic sense.[4] Note that in this model the state variable $x_n \approx U_0(\varepsilon n)$ changes slowly relative to the noise variable y_n.

These calculations demonstrate two interesting uses of calculus to approximate solutions of difference equations. We use the results in the following simulations.

2. Simulation and Data Mining

How would data-mining work on such a randomly perturbed problem? This might begin with simulating the model (Monte Carlo simulations), and then searching the resulting database for various structures. The following example shows how the search can be guided by analysis.

To fix ideas, we consider a problem from population genetics where u_n describes a proportion of the gene pool of type A in a single locus, two allele genetic trait having alleles A and B.

The Fisher-Wright-Haldane (FWH) model[2] for this is

$$u_{n+1} = \frac{r_n u_n^2 + s_n u_n(1-u_n)}{r_n u_n^2 + 2s_n u_n(1-u_n) + t_n(1-u_n)^2}.$$

In the slow selection case, all of the fitnesses are almost equal, but the perturbations are random:

$$\begin{aligned} r_n &= r + \varepsilon\rho(y_n) \\ s_n &= r + \varepsilon\sigma(y_n) \\ t_n &= r + \varepsilon\tau(y_n) \end{aligned}$$

where $\{y_n\}$ is a stochastic process (perhaps a vector of them) as before.

We define the averages of the data over Y to be

$$\begin{aligned} \bar\rho &= \int_Y \rho(y)\, m(dy) \\ \bar\sigma &= \int_Y \sigma(y)\, m(dy) \\ \bar\tau &= \int_Y \tau(y)\, m(dy). \end{aligned}$$

Then we use the result above to show[4] that

$$\frac{dU_0}{ds} = U_0(1-U_0)\left(U_0(\bar\rho - 2\bar\sigma + \bar\tau) + (\bar\sigma - \bar\tau)\right).$$

The following two examples show two cases: In the first, there is a stable polymorphic state. In the second, A wins.

The model is simulated 500 times for $n = 1, \ldots, 1000$. The Monte Carlo data base generated by this is shown in Fig. 1. A cursory look does not reveal much structure. However, analysis showed that the solution should be an average value $(U_0(n))$ plus a diffusion process, so we plot a histogram of the solution values for each n to describe the distribution of solutions at that time.

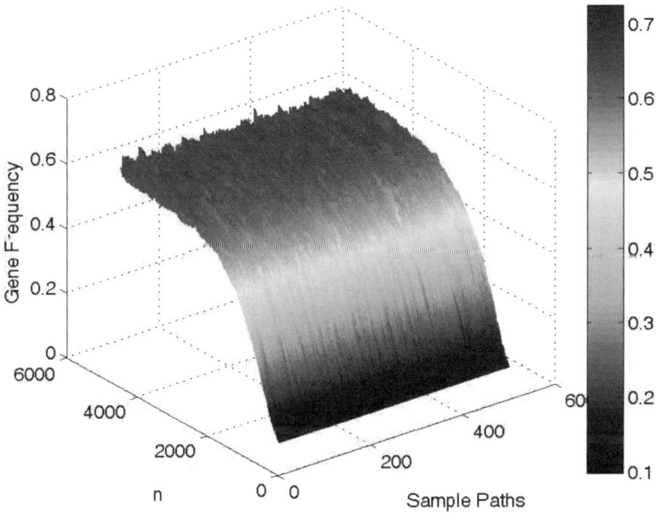

Fig. 1. Monte Carlo Data Base for random perturbations of the FWH model.

The first example, called heterosis, has $\bar{\tau} < \bar{\rho} < \bar{\sigma}$. The initial and final distributions are plotted in Fig. 2. The distribution is bell-shaped and centered at $(\bar{\tau} - \bar{\sigma})/(\bar{\rho} - 2\bar{\sigma} + \bar{\tau})$, and it describes a balance between the selective forces moving the populations toward the equilibrium of the averaged system and the drift introduced by random perturbations of the fitnesses.

Next, we simulate the case where $\bar{\rho} > \bar{\sigma} > \bar{\tau}$, and we use the same initial conditions (all at 0.1). As shown in Fig. 3, A wins, and the final distribution reflects Wright's formula showing how selective pressure pushing populations to the right is balanced by random drift.[5]

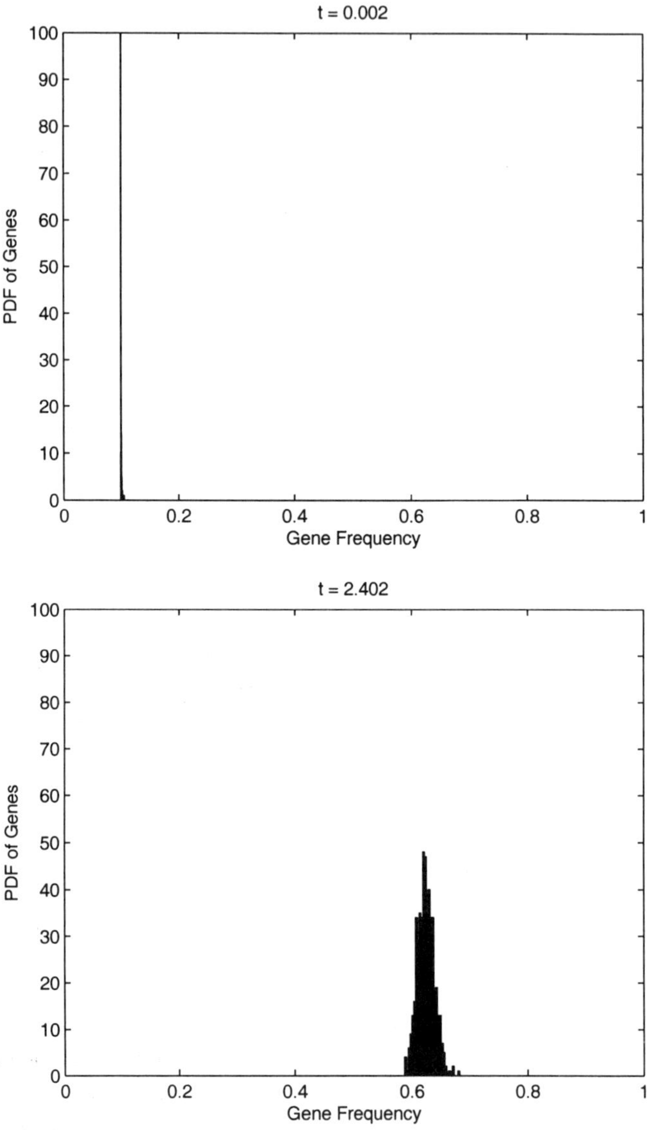

Fig. 2. Initial (top) and Final (bottom) distributions in the case of heterosis.

3. Volterra Integral Equations

The following examples illustrate problems for simulation and data mining in some non-Markovian models. Consider a Volterra integral equation for

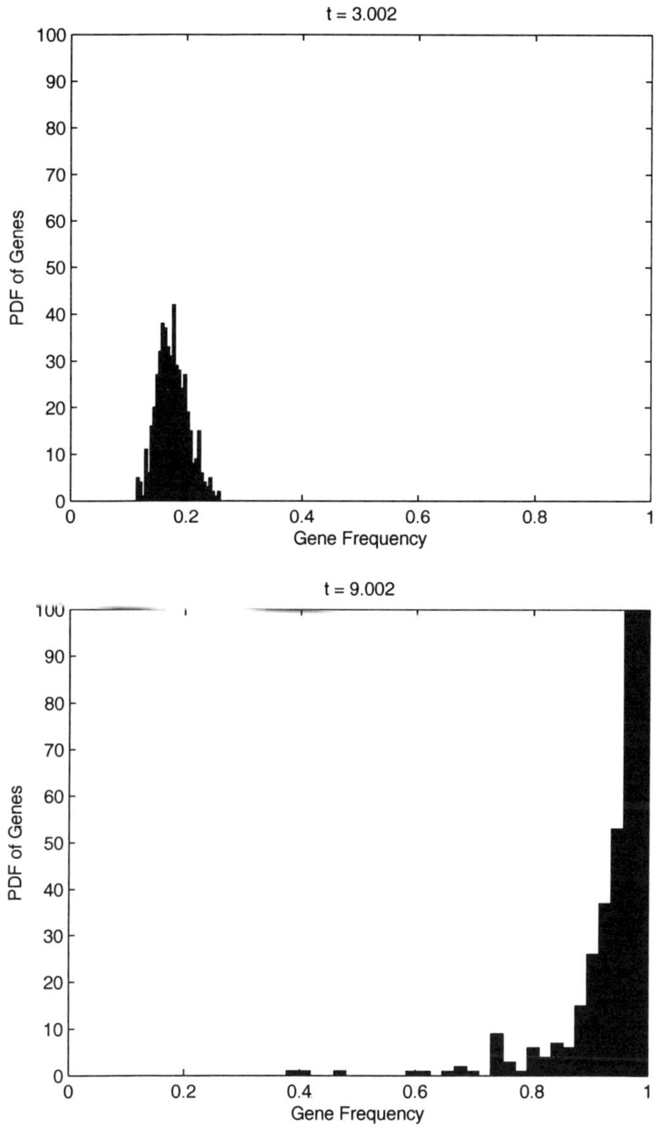

Fig. 3. Distribution of solutions at $t = 3$ (top) and at equilibrium (bottom).

$x(t)$:

$$x(t) = \phi(t) + \int_0^t K\left(t, t', y(t'/\varepsilon), x(t')\right) \, d\mu(t')$$

where

- $x(t)$ is a vector of state variables for each $t \in [0, \infty)$,
- $\phi(t)$ is given and is dominated by a decaying exponential,
- y is an ergodic random process in a compact space Y (it is strongly mixing, and stationary or homogeneous Markov) having ergodic measure m,
- the kernel $K(t, t', y, x)$ is a smooth function of t, t', x and a measurable function of $y \in Y$,
- μ is a Stieltjes measure, which weights contributions from previous times.

We first average the system:

$$\bar{x}(t) = \phi(t) + \int_0^t \bar{K}\left(t, t', \bar{x}(t')\right) d\mu(t')$$

where

$$\bar{K} = \int_Y K(t, t', y, x)\, m(dy).$$

Then we show[4] that the solution of the original problem satisfies

$$\frac{x(t, \varepsilon) - \bar{x}(t)}{\sqrt{\varepsilon}} \approx x_1$$

where x_1 is a stochastic process determined by solving an equation of the form

$$x_1(t) = z(t) + \int_0^t L(t, t')\, x_1(t')\, d\mu(t')$$

where z is a Gaussian process and L is a kernel, both of which can be calculated. Since the resolvent of this equation is a linear operation on z, x_1 is a Gaussian process as well.

The simulation/data-mining approach can have great difficulty with such problems. The following example illustrates one problem.

We consider the case where

$$\bar{K}(t, t', x(t'))\, d\mu(t') = \frac{1}{\delta} \chi_{[T-\delta, T]}(t - t') f(x(t')) dt'$$

The averaged equation becomes

$$\begin{aligned} \bar{x}(t) &= \phi(t) + \int_0^t \frac{1}{\delta} \chi_{[T-\delta,T]}(t - t') f(\bar{x}(t'))\, dt' \\ &= \phi(t) + \frac{1}{\delta} \int_{t-T}^{t-T+\delta} f(\bar{x}(t'))\, dt' \\ &\approx \phi(t) + f(\bar{x}(t - T)) + O(\delta). \end{aligned}$$

We analyze this using straightforward iteration.

Let $t = nT$ and $x_n = \bar{x}(nT)$. Then, since ϕ dies out, we eventually have

$$x_{n+1} = f(x_n).$$

To fix ideas, we consider Ricker's model where

$$f(x) = r\,x\,e^{-x}.$$

We simulate this iteration by picking an initial point, say $x_0 = 1.0$, then iterating the function to get samples of the solution $\{x_0, x_1, \ldots, x_N\}$. Two cases are illustrated in Fig. 4.

We construct a histogram of these values to detect invariant structures in the solution. We do this for many values of $r \in (0, 20)$,[1] and plot the iterate density histogram of $x_n e/r$ for 400 values of $r \in (0, 20)$ in Fig. 5.

The problem in this case is that the averaged equation itself has chaotic behavior in its solutions, so we cannot expect any finite perturbation to accurately reflect the solution's behavior, since it will involve a combination of the effects of the random perturbations mixed with the underlying chaotic dynamics.

A second problem for simulation of such systems is that solving even the averaged integral equation can pose difficult numerical problems.[6]

4. Conclusions

Simulation/data-mining can be very powerful method for gaining understanding of complex systems that are posed as mathematical models. Mathematical analysis can guide interpretations of simulations.

However, accurate simulation is not always possible, and may pose challenging numerical analysis problems.

Iteration between ordinary language models (precise descriptions of experiments and their outcomes) and mathematical language models (canonical models, analysis, simulation and visualization), continue to be essential to understanding physical and biological systems. Simulation/data-mining plays an important role in this process.

Non-Markovian models pose considerable computational challenges, and they are important in many areas of science and engineering; e.g., brain science (polychronization), population dynamics (renewal, epidemiology), electronics (filter design), materials (viscoelastics), and control (feedback delays). We should include study of these methodologies (for example, Volterra Integral Equations) in our curriculum; these have not yet fallen behind the computer screen.

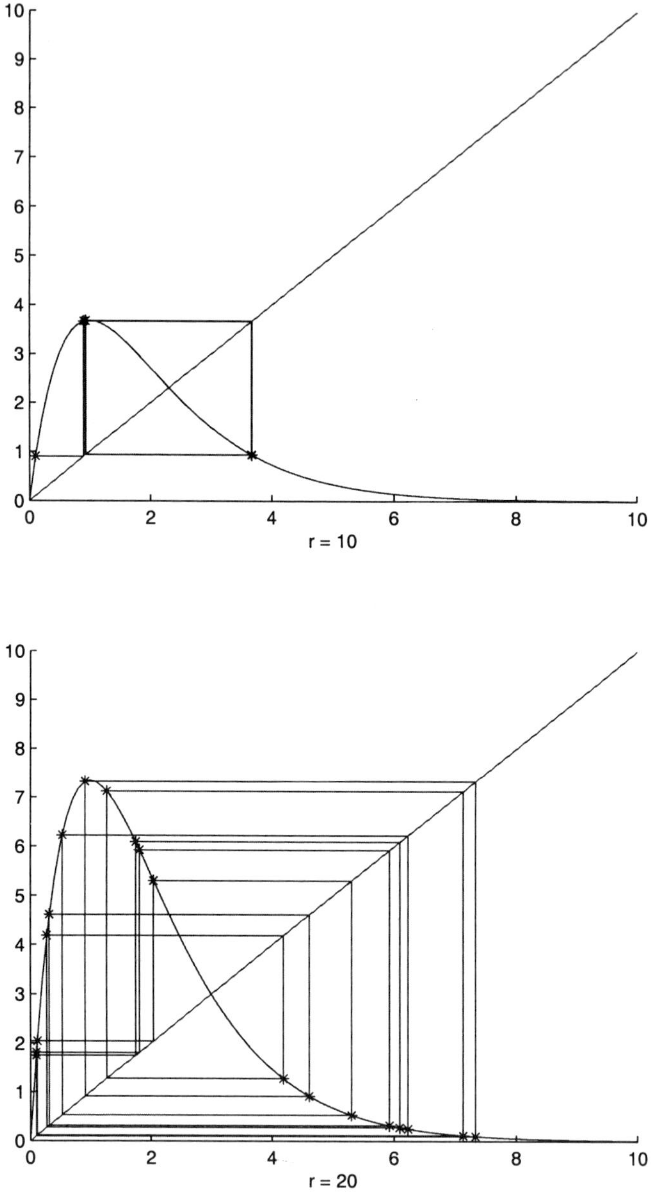

Fig. 4. Geometric iteration of Ricker's Model: r=10 (top) and r = 20 (bottom)

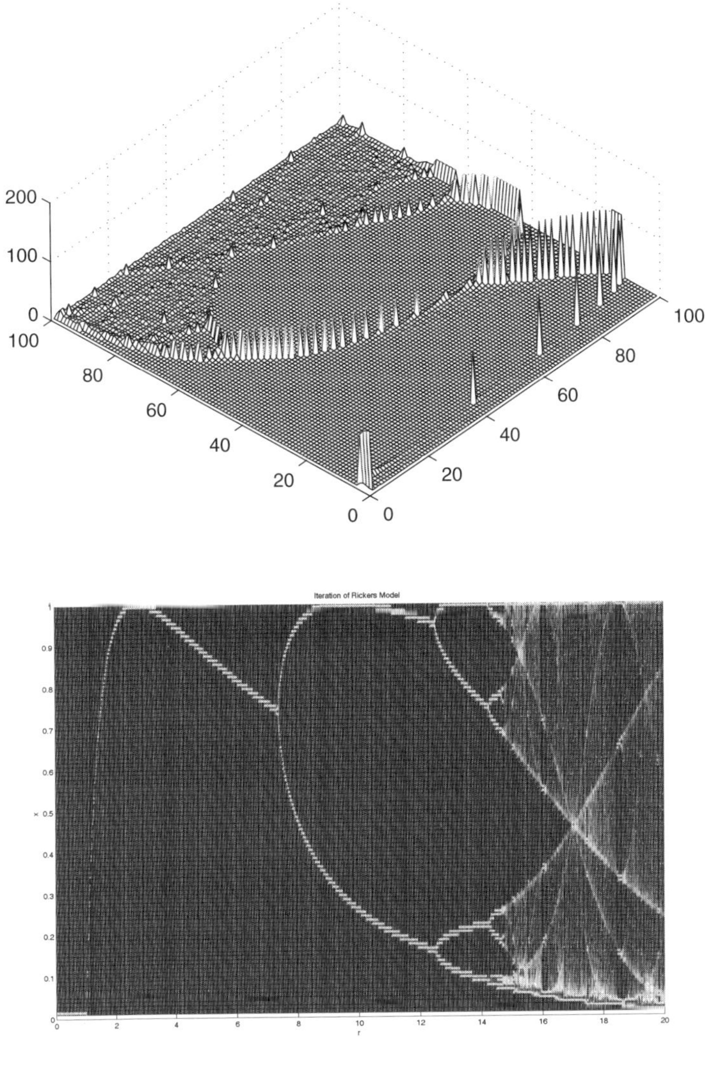

Fig. 5. Ricker's Model: Iterate Density Distributions:[1] Top: Elevation, Bottom: Top view.

References

1. F.C. Hoppensteadt, J.M. Hyman, Periodic solutions to a logistic difference equation, SIAM J. Appl. Math. 32 (1977) 73-81.

2. F.C. Hoppensteadt, Mathematical Methods of Population Biology, Cambridge University Press, 1982.
3. F.C. Hoppensteadt, Analysis and Simulation of Chaotic Systems, 2nd ed., Springer, New York, 2000, p. 207.
4. A. Skorokhod, F.C. Hoppensteadt, H. Salehi, Random Perturbation Methods with Applications in Science and Engineering, Springer, New York, 2002
5. D. Ludwig, Stochastic Population Theories, Springer-Verlag, 1972.
6. B.Zubik-Kawal, Z. Jackiewicz, F.C. Hoppensteadt, Numerical solution of Volterra integral and integro-differential equations with rapidly vanishing convolution kernels, BIT, 47 (2007), no. 2, 325–350.

WHITHER BIOSTOCHASTICS IN COMPUTATIONAL BIOLOGY AND BIOINFORMATICS

PRANAB K. SEN

Department of Statistics, University of North Carolina
Chapel Hill, NC 27599, USA
E-mail: pksen@bios.unc.edu

Principles of molecular biology, (multi-factorial) genetics, genomics, pharmacogenomics and toxicogenomics govern computational biology and disinformation, in general. There is ample room for biostochastics to comprehend the basic difference between mathematical determinacy (prevailing in physical and engineering sciences) and biological diversity as commonly perceived in this emerging interdisciplinary field of research. Various disciplines, including biology, computational mathematics, computer science, physics and statistics, are looking through the edges of this prism in their own way. With the evolution of information technology, computational facilities are expanding in an astounding way, and thus providing access to massive data sets arising in disinformation and genomics. Knowledge discovery and data mining tools are increasingly advocated for quantitative assessments, albeit often without adequate statistical validation and interpretation. High dimensionality relative to low sample size restraints create impasses for adoption of conventional statistical methodology in genomic studies and require highly nonstandard statistical appraisals. Biostochastic undercurrents deserve a critical assessment in such complex studies. Beyond parametrics perspectives are highlighted in this appraisal.

Keywords: biostochastics, disinformation, data models, genomics

1. Introduction

There are three quantitative approaches running in parallel: (1) computational mathematics, (2) computational biology, and (3) computational statistics, albeit none alone can meet the need of current day interdisciplinary research. Principles of molecular biology, (multi-factorial) genetics, genomics, proteomics, pharmacogenomics and toxicogenomics govern computational biology, system biology and disinformation. Bioinformatics, flooded with massive datasets they generate in furious details and astounding pace, are appealing increasingly to *knowledge discovery and data mining (KDDM)* or *statistical learning* for procuring ready-made quantitative as-

sessments. Faced with the basic objectives of mapping the *disease genens* and assessing the *gene-environment interactions*, such computational mathematical tools often bypass statistical challenges posed by some thirty thousand genes whose (clustered) interaction with diseases and environmental stressors are yet to be completely statistically validated and interpreted. Excessive cost of scientific studies often precludes good statistical design and routine adaption of standard statistical or biomathematical tools; stochastic undercurrents deserve critical appraisal prior to making any quantitative assessment. Some of these biostochastic perspectives in such exceedingly *high dimensional low sample size data* (HDLSS) models are appraised along with some illustrations from microarray and SNP data models. Beyond parametrics perspectives are highlighted in this appraisal. What is biostochastics? How does it differ from biometry and biostatistics? In what respect disinformation is different from conventional computational statistics or informatics? We may use the following terminology.
Biostochastics: Biologically interpretable stochastic modeling and analysis (i.e., stochastics) in very large biological, including genomics, data models; it includes thereby statistical justification and interpretation of KDDM and statistical learning, and annexes theoretical (mathematical) biology and biomathematics in this general setups.
Bioinformatics: Application of computational mathematics, computational (bio-)statistics and computer intensive information technology to quantitative assessment of very large biological systems, genetic and genomic data models.
Unlike in mathematical sciences, in biological systems, due to bio-diversity (genetic variation), neither exact mathematical laws are yet to be laid down precisely nor they are expected to exist. Rather, computational mathematical tools are primarily used to guess general patterns in a quantitative setup without eliminating the possibility of incorrect assessments. Coordination of sustainable biological interpretation with valid and efficient statistical conclusion makes biostochastics different from conventional biostatistics. Yet, there are many interesting features in mathematical and biostatistics developments that are adaptable in HDLSS setups in disinformation / biostochastics, and I would like to illustrate some of these statistical developments.

In order to outline the transition from biometry to biostatistics to biostochastics, in Section 2, we briefly outline the genesis of biostochastics through the developments in clinical trials and medical studies. We also stress the difference between clinical trials and disinformation setups there. Section 3 is devoted to HDLSS models where biostochastics play a crucial role with an illustration from a microarray data model. Section 4 is devoted to a SNP data model setup. The concluding section stresses the importance of multiple hypotheses testing and multiple comparisons procedures in HDLSS setups. In this highly nonstandard situation, the Chen-Stein (1975) Poisson approximation theorem is contrasted with the classical Ballot (under the pseudo-name of Simes (1986)) theorem and their roles are highlighted in this context.

2. Genesis of Biostochastics

As a close ally to the developing discipline of statistics, some 75 years ago, biometry captured the attention of researchers working in biological, agricultural and genetics fields. No wonder in those early days, the same rationality prevailing in statistical analysis was surging in biometric research as well. However, it did not take long to comprehend the basic differences between pure mathematical statistics and biometric perspectives. Such biometric studies were mostly on biological units other than human being as subjects so that consideration of implementation of *randomization, replication* and *local control*, Fisher's (1932) pioneering measures underlying design of experiments, could be justified to a greater extent. There were differences in the mode of collection of sample observations, justification of equal probability sampling, independence of observations and all sorts of allied things that initiated the development of biostatistics about 60 years ago. Another important factor in this development was the use of observational studies instead of planned experiments that required somewhat different methodology. Transformations of variates to accommodate plausible underlying distributions were advocated for greater flexibility. From linear models to nonlinear models transitions were welcome to capture more insight into statistical foundations. Yet, in this biostatistics development, statistical rationality was quite in concordance with classical statistics. Compared to biometry, biostatistics made some significant efforts to address more appropriately some of the intricate statistical problems arising in epidemiology, nutrition, health management, environmental health, and of course, deeper into medical studies as well as clinical sciences. Apart from emphasis on computational aspects, inducing developments of various statistical packages, there has been a basic factor: *Biostatistics is more of an art in understanding the basic statistical undercurrents of specific problems, translating them to interpretable statistical problems, and only then implementing statistical tools for drawing statistical conclusions.* During the 1960s and onwards, biostatistics became more viable in all health sciences research endeavors, including chronic diseases and disorders, clinical trials and drug developments. This phase continues even now, although it has become far more challenging due to the intervention of disinformation. Let us examine this transition in the light of statistical methodology pertaining to such nonstandard problems. Stochastic models for epidemics and chronic diseases and disorders have incorporated some innovative stochastic processes leading to novel statistical tools and analysis procedures. Martingale characterizations and (semi-) Markov processes are specially noteworthy in this context. They are also useful illustrations of the effective role of probability theory and theoretical statistics in such fruitful applications and related biostatistics methodology. In a randomized clinical trial set for a placebo and treatment group, the ordered outcome variables are to be recorded along with the baseline explanatory variables and possibly other covariates. The same batch of subjects are followed over a duration of time, thus inviting censoring as well as nonindependence of the observed responses. In this time-sequential setup, the basic structure of independent and homogeneous increments may not generally hold, and as a result classical statistical inference tools, including sequential ones, may not be applicable. Moreover, *interim analysis* allowing to have multiple looks into the accumulating data sets leads to what are known as *multiple hypotheses testing* (MST) and *repeated significance testing* (RST), without assuming independence and homogeneity of increments. Still, in this setup, the number of 'looks' is generally small compared to the sample size(s). That makes it possible to modify existing statistical tools to suit such applications. Martingale methods based statistical procedures have been extensively developed to suit such applications (viz., Sen 1981).

To emphasize the evolution of biostochastics, at this stage, let us examine the impact of dosimetry and environmental health sciences on clinical trials and biomedical research. In environmental health studies, toxicology occupies a focal stand. Generally, exposure to environmental toxicants are prolonged but at low levels. However, human subjects can not be generally used to conduct such studies, and subhuman primates are used as surrogates. Further, from cost and feasibility considerations, for such subhuman primates, the exposure level is increased and the duration of the study is made much shorter. This issue raises some broad questions: Can the total exposure, i.e., time-duration times the exposure level, be taken as an effective measure of the toxicity intake? Can the exposure level be kept at a constant mark through out the study? How drug-addiction factors can distort such simpler models? More importantly, how to extrapolate from stochastics (in metabolism and drug response) from mice to man? For human being, the aging aspects, changing environmental setups and geopolitical factors, often interacting intensively with mental traits and attitude towards life, make it quite difficult to conceive of a laboratory setup for such environmental health studies. Human metabolism and drug-response pattern may be highly different from subhuman primates. Modern dosimetric studies aspire to probe into such differential aspects, although more thoroughly with biological and chemical studies, and there is a genuine need for statistical reasoning, modeling and analysis protocols in this interdisciplinary field of research. There is a basic difference between animal studies (or dosimetry) and clinical trials, involving human subjects, namely, controlled laboratory setups to largely uncontrolled environment wherein an enormous disparity in physical as well as metabolic characteristics and other epidemiologic end-points can mar the simplicity of statistical modeling to a greater extent. In the past, clinical trials have mostly been conducted with studies of *symptomatic effects* whereas the actual objectives could have been *disease-disorder detection* and cure. Drug developers, pharmaceutical groups and regulatory agencies, for practical reasons, used to focus on treatments to relieve symptoms, albeit without any guarantee of adequately matching the treatment objectives. In that way, conventional *placebo-controlled trials* (PCT) have been advocated knowing well that such trials could be inappropriate on medical ethics and affordability considerations. The past fifteen years have witnessed a steady flow of research on various approaches (including the so called *active controlled equivalence trials* (ACET) and *randomized dynamic treatment regimen trials*) to eliminate some of the drawbacks of PCT. Nevertheless, they are associated with an elevated level of complexity of statistical modeling and analysis prospectives. *Constrained statistical inference* (CSI) procedures (Silvapulle and Sen 2004) has greater scope in this respect, albeit in many cases they may need larger sample sizes - a restraint that may not be generally tenable in many clinical trials and medical studies.

Let us appraise the burden of disinformation in clinical trials and toxicological studies, and trace the evolution of biostochastics in this context. With the ongoing evolution of disinformation, clinical trials and toxicological studies are encountering some challenges. The surging emphasis on pharmacogenomics, underlying drug responses, detection of *disease genes* and *gene-environment interaction*, has created a tremendous statistical task which will be elaborated in the next section. Basically, there are some thousands of genes acting interdependently and regulating our metabolism, health and biological functioning, and life as a whole. Yet the genes have possibly different roles and possibly different levels of interactions with specific diseases, disorders and environmental stressors. Clinical trials are now charged with finding the genes causally or statistically associated with specific diseases (or disorders) as well as their *pharmacokinetics* and *pharmacodynamics* with specific drug development. This is a distinct shift from symptomatic effects to genetic/genomic effects study objectives. Instead of clinical trials with human subjects, it calls for additional refinements: *microarray* and *proteomics* studies at the molecular level with tissues and cells. Indeed, biotechnology innovations permit such experimental studies, albeit at a huge expense, and thus resulting in a far smaller sample size, This is the central point of this study: The extraction of statistical information from large biological systems, particularly, from genomics studies. Standard (bio-)statistical tools are of little use in this context, and the development of biostochastics aims to capture the novel approaches in this perspective.

3. HDLSS Paradigms

The scenario in clinical trials involves accumulating data in a follow-up scheme where conventional independent and homogeneous increments assumption may not typically hold. Nevertheless, interim analysis generally rests on a finite number of tests on accumulating datasets where martingale theory can be used with advantage. In HDLSS data models, such martingale characterizations are usually not tenable. Even other conventional asymptotics are of little use. As an illustration, consider a genomic model arising in microarray data analysis. The microarray technology allows simultaneous studies of typically thousands of genes (K), possibly differentially expressed under diverse biological or experimental setups, with only a few (n) arrays. Such experiments are generally excessively costly, thus preempting the possibility of having a large number of arrays, and thereby, resulting in the $K >> n$ environment. We may refer to Lobenhofer et al. (2002) where for a set of 1900 genes, arranged in rows, the gene expressions were recorded at 6 time points, with 8 observations at each time point. Thus virtually $K = 1900$ and $n = 48$, clearly signaling the $K >> n$ environment. The gene-expression levels are measured by their color intensity (or luminosity) as a quantitative (nonnegative) variable, either on the (0, 1) or 0 - 100 per cent scale, or (based on the log-scale) on the real line $\mathbf{R}$. A gene associated causally or statistically) with a target disease is known as *disease gene* (DG) while the others as *nondisease genes* (NDG). Gene expression levels under different environment cast light on plausible *gene-environment inter-*

action (or association) so that if the arrays are properly designed, *mapping disease genes* may be facilitated with such microarray studies. One of the main issues is identifying differentially expressed genes among thousands of genes, tested simultaneously, across experimental conditions. Typically, for a target disease, there are only a few DG while the NDG comprise the vast majority. A NDG is expected to have a low gene expression level while a DG is expected to have generally higher expression levels. Thus, a natural *stochastic ordering* of gene expression levels of the DG with varying disease severity is plausible while the NDG expression levels are expected to be stochastically unaffected by such disease level differentials. Microarray data go thorough a lot of standardization and normalization so that conventional simple models, such as the classical MANOVA models may rarely be totally adaptable. The usual requirement of the positive definiteness of the sample covariance matrix is not satisfied in the $K >> n$ environment so that the likelihood based MANOVA procedures fail to operate (Sen 2006). If the arrays are indexed by an explanatory or design variate (t) that possesses an ordering (not necessarily linear), then the stochastic ordering could be exploited through suitable nonparametric techniques. The main difficulty in modeling and statistically analyzing microarray data stems out of the huge dimensionality (K) of the genes compared to the number of arrays. While the different arrays may sometimes be taken to be at least statistically independent, the genes may not be so. Moreover, not so much is known about the spatial topology of the genes, and even their genetic distances. This makes it difficult to incorporate functional regression models in a nonparametric setup; the paucity of n (compared to K) being a stumbling roadblock for such smoothing tools based nonparametrics. There is another factor that merits our attention. The gene expression levels for the different genes in an array are neither expected to be stochastically independent nor (marginally) identically distributed. Sans such an i.i.d. clause, standard parametrics typically adaptable for fMRI models (albeit mostly done in a Bayesian coating) may encounter roadblocks for fruitful adaptation in microarray data models. Thus, structurally, such data models are different from those usually encountered in nonparametric functional regression models. Although, there is a multi-sample or time-series flavor in the setup of the different arrays, the main complication in statistical modeling arises due to the vast number of genes and an anticipated spatial dependence pattern without a parametric perspective. For this reason, a pseudo-marginal approach (Sen 2008) is highlighted here. This approach exploits the marginal nonparametrics fully and renders some useful modeling and analysis convenience.

For a set of n arrays (sample observations) with a design variate t_i associated with the ith array, for $i = 1, \ldots, n$, we assume that the t_i are ordered, i.e., $t_1 \leq t_2 \leq \cdots \leq t_n$, with at least one strict inequality. We do not, however, impose any linear or specific parametric ordering of these design variates. The multisample (ordered alternative) model is a particular case where n can be partitioned into I subsets of sizes $n_1, \ldots, n_I$ such that within each subgroup, the t_i are the same while they are ordered over the I different subsets. For the ith array, corresponding to the K genes (positions), we have a gene expression level X_{ik}, $k = 1, \ldots, K$, resulting in K-vectors $\mathbf{X}_i = (X_{i1}, \ldots, X_{iK})'$, for $i = 1, \ldots, n$. The joint distribution function of $\mathbf{X}_i$ is denoted by $F_i(\mathbf{x})$, $\mathbf{x} \in \mathbf{R}^K$. Further, for X_{ik}, the marginal distribution is denoted by $F_{ik}(x)$, $x \in \mathbf{R}$, for $k = 1, \ldots, K$; $i = 1, \ldots, n$. For a given i, the F_{ik}, $k = 1, \ldots, K$ may not be generally the same, and moreover, the X_{ik}, $k = 1, \ldots, K$ may not be all stochastically independent.

If a gene k is NDG and the t_i reflect the variability of the disease level, then the $F_{ik}, i = 1, \ldots, n$ should be the same. On the other hand, for a DG k, for $i < i'$, X_{ik} should be stochastically smaller than $X_{i'k}$ in the sense that the $F_{ik}, i = 1, \ldots, n$ should have the ordering

$$F_{1k}(x) \geq F_{2k}(x) \geq \cdots \geq F_{nk}(x), \forall\ x \in \mathbf{R}.$$

Therefore, we could have a characterization of DG and NDG based on the following stochastic ordering: For a NDG k, the F_{ik}, $i = 1, \ldots, n$ are all the same, this being denoted by the null hypothesis H_{0k}, while for a DG k, the stochastic ordering holds which we denote by H_{1k}, for $k = 1, \ldots, K$. In this marginal formulation, we have a set of K hypotheses corresponding to the K genes, and whatever appropriate test statistic (say T_{nk}) we use for testing H_{0k} vs. H_{1k}, these statistics may not be, generally, stochastically independent. The basic problem is therefore to test simultaneously for

$$H_0 = \cap_{k=1}^K H_{0k} \text{ vs } H_1 = \cup_{k=1}^K H_{1k}, \tag{1}$$

without ignoring possible dependence of the tests statistics for the component hypotheses testing H_{0k} vs H_{1k}, for $k = 1, \ldots, K$. This makes it appealing to follow the general guidelines of the Roy (1953) *union-intersection principle* (UIP), albeit in a marginalization (i.e., adapting a finite union and finite intersection scheme), and thus permitting a more general framework so as to allow simultaneous testing and classification into DG / NDG groups. Our approach is based on the classical Kendall tau statistics for each of the K genes and then incorporate these (possibly dependent) marginal statistics in a composite scheme for classification. For the kth gene, based on the n observations X_{ik}, $i = 1, \ldots, n$, and the tagging variables $t_1, \ldots, t_n$, we define the Kendall tau statistic as

$$T_{nk} = \binom{n}{2}^{-1} \sum_{1 \le i < i' \le n} sign(X_{i'k} - X_{ik}) sign(t_{i'} - t_i), \tag{2}$$

for $k = 1, \ldots, K$. Conventionally, we take $sign(0) = 0$. Note that T_{nk} is a (generalized) U-statistic of degree 2 (Hoeffding 1948). Further, note that by (3.1), we may set $\mathcal{S} = \{(i, i') : t_i < t_{i'}; 1 \le i < i' \le n\}$ and let N be the cardinality of the set $\mathcal{S}$. Then by (3.1), $n - 1 \le N \le \binom{n}{2}$. Moreover, we may rewrite T_{nk} as

$$T_{nk} = \binom{n}{2}^{-1} \sum_{\mathcal{S}} sign(X_{i'k} - X_{ik}), \; k = 1, \ldots, K, \tag{3}$$

where the set $\mathcal{S}$ depends on the ordering of the t_j and thereby remains the same for every $k = 1, \ldots, K$. Note further that whenever $N < \binom{n}{2}$, the range of variation of T_{nk} is $(\, -N/\binom{n}{2}, N/\binom{n}{2} \,)$ which is contained in the interval $(-1, 1)$. That is why we rescale it as

$$T_{nk} = N^{-1} \sum_{\mathcal{S}} sign(X_{i'k} - X_{ik}), \tag{4}$$

whose range is exactly $(-1, 1)$, albeit the distribution being still discrete. Note that for any $k (= 1, \ldots, K)$, under H_{0k}, for every $i \ne i'$, the difference $X_{i'k} - X_{ik}$ is symmetrically distributed around 0, and hence, $E_{0k}\{sign(X_{i'k} - X_{ik})\} = 0$ so that

$$E_{0k}\{T_{nk}\} = 0, \; \forall \; k = 1, \ldots, K. \tag{5}$$

Further, the marginal distribution of T_{nk} under H_{0k} is generated by the $n!$ equally likely permutations of the X_{ik} among themselves. Therefore when all the F_{ik} are continuous, ties among the observations being negligible with probability 1, T_{nk} is distribution-free under H_{0k}. This distribution may depend on the set $\mathcal{S}$ but that being the same for all $k (= 1, \ldots, K)$, we conclude that under H_0, marginally each T_{nk} is distribution-free and these K statistics all have the same marginal distribution. For large sample sizes asymptotic normality results are well known (Hoeffding, 1948). In our setup, perhaps the exact permutation distribution plays a greater role, as will be illustrated later on.

The behavior of T_{nk} under alternatives would naturally depend on their stochastic ordering and these statistics will not be exact distribution-free nor possibly have identical marginal laws. Nevertheless, under H_{0k}, for every $i < i'$, $X_{i'k} - X_{ik}$ has a distribution tilted to the right, so that

$$E\{T_{nk} \mid H_{1k}\} \geq 0, \ \forall\, k = 1, \dots, K. \tag{6}$$

This motivates us to use tests based on the marginal statistics T^o_{nk} using the right hand side critical region, or equivalently the right-hand sided p-values. Recall that the distribution of each T_{nk}, at least for n not too large, is discrete, but that is not going to be of any particular concern. A greater concern is to incorporate possible stochastic dependence among the K statistics T_{nk}, $k = 1, \dots, K$ (even under the null hypothesis) and their possible heterogeneity when some of the H_{1k} are true. This concern is even more acute in our contemplated $K >> n$ environment. A basic problem is to formulate suitable multiple hypotheses testing procedures in order to assess which hypotheses are to be rejected subject to a suitably defined Type I error rate. This is outlined in Section 5.

4. HDLSS for Qualitative Data Models

As an illustration consider a DNA nucleotide or RNA protein model: There are usually a large number (K) of genes, and at each position, the response variable is categorical with C possible (unordered) outcomes, indexed as $c = 1, \dots, C$; $C = 4$ for nucleotide data and 20 for RNA data models. Thus, the probability law is defined on a $C^K - 1$ simplex, requiring for the applicability of Categorical MANOVA procedures that $n >> C^K$ which contradicts the $K >> n$ environment. This is the curse of dimensionality problem in qualitative genomics. As in the previous section, we consider a multi-sample problem with I groups where for each group we have a probability law $\boldsymbol{\pi}_i, i = 1, ..., I$. The observable sample random vectors are $\mathbf{X}_{ij} = (X_{ij1}, \dots, X_{ijK})$ with X_{ijk} taking on the indexes $1, \dots, C$ according as the jth observation in the ith group belongs to cell indexed as $1, ..., C$. For any pair of observations $\mathbf{X}$ and $\mathbf{Y}$, we define the Hamming distance

$$d(\mathbf{X}, \mathbf{Y}) = K^{-1} \sum_{k=1}^{K} I(X_k \neq Y_k). \tag{7}$$

If, for the ith group, kth position, we denote the multinomial cell probability vector by $\boldsymbol{\pi}_{ik} = (\pi_{ik1}, \dots, \pi_{ikC})'$, then $E\{I(X_{i1k} \neq X_{i2k}\} = 1 - P\{X_{i1k} = X_{i2k}$, so that $E\{d(\mathbf{X}_{ijk}, X_{ij'k})\} = 1 - \boldsymbol{\pi}_{ik}{}' \boldsymbol{\pi}_{ik}$. Thus, if $\mathcal{H}_i = 1 - K^{-1} \sum_{k=1}^{K} \boldsymbol{\pi}'_{ik} \boldsymbol{\pi}_{ik}, i = 1, \dots, I$ stand for the Hamming distance for the I groups, we have

$$D_{ii} = K^{-1}\binom{n_i}{2}^{-1}\sum_{k=1}^{K}\sum_{1\leq j<j'\leq n_i} I(X_{ijk} \neq X_{ij'k}) \tag{8}$$

the sample Hamming distance whose expectation is $\mathcal{H}_i$, for $i = 1, \ldots, I$. These are the arithmetic means of the K gene-wise Gini-Simpson indexes. Likewise, for any pair $(i \neq i')$,

$$\mathcal{H}_{ii'} = K^{-1}\sum_{k=1}^{K} P\{X_{ijk} \neq X_{i'j'k}\} \tag{9}$$

is the Hamming distance between the ith and i'th groups. Their sample counterparts are

$$D_{ii'} = K^{-1}n_i^{-1}n_{i'}^{-1}\sum_{k=1}^{K}\sum_{j=1}^{n_i}\sum_{l=1}^{n_{i'}} I(X_{ijk} \neq X_{i'lk}). \tag{10}$$

Along with a subgroup decomposability property of the pooled Hamming distance into within group and between group components (resembling the classical ANOVA decomposition) the sample Hamming distances were used to provide suitable tests (Pinheiro et al. 2005, Sen 2006, Sen et al. 2007). In this study, along the lines of the preceding section, we use the coordinatewise Gini-Simpson indexes and formulate in Section 5 some alternative tests. Note that the coordinatewise Gini-Simpson indexes are

$$G_{iik} = \binom{n_i}{2}^{-1}\sum_{1\leq j<j'\leq n_i} I(X_{ijk} \neq X_{ij'k}),\ k = 1, \ldots, K. \tag{11}$$

We also define the between group Gini-Simpson indexes as $G_{ii'k}$, for $1 \leq i < i' \leq I$, and based on these coordinatewise statistics, consider a test statistic T_{nk} for $k = 1, \ldots, K$. Whereas in the case of the Kendall tau statistics, considered in Section 3, we have an exact distribution-free test for each of the K coordinates, here the null hypothesis distribution generally depend on the unknown $\boldsymbol{\pi}_i$ and hence the tests are not exact distribution-free. Nevertheless, the same permutation principle holds for this case also, and hence, that can be used to generate permutational null distribution of the T_{nk} (Sen et al. 2007). As such, we present the multiple hypothesis testing problem in a common vein for both the quantitative and qualitative models.

5. Adaptation of the Chen-Stein Theorem

For the microarray data model, taking into account plausible inter-gene stochastic dependence and heterogeneity, we need to prescribe statistical modeling and analysis tools incorporating *dimensional asymptotics* where K is made to increase indefinitely while n, being small compared to K, may or may not be adequately large. For the qualitative data model too, a similar situation arises, and hence, we present the following approach to formulate this multiple hypothesis testing problem.

In order to control the so called *family wise error rate* (FWER), say to a preassigned level $\alpha : 0 < \alpha < 1$, with possibly K testing components, for individual tests, effectively we have a level of significance α/K, very small for large K. The exact null distribution for small $n_i, i = 1, \ldots, I$ is discrete and hence the associated P-values (i.e., observed significance levels) may not have the uniform (0,1) distribution. However, we can use the permutation distribution to get these P-values. Even so, we need the mass points near the lower end-point (0). The crux of the problem is therefore to determine such a critical level c_α. The joint distribution of the $T_{nk}, 1 \leq k \leq K$, even under the null hypothesis H_0, depends on the underlying K-dimensional distribution F_i so that the usual technique of finding out the critical level of T_n^* from this joint distribution may be intractable. The permutation distribution over the possible $n!/\prod_{i=1}^{I} n_i!$ possible partitioning of the I groups of course provide a general idea of this joint distribution but for small n it fails to be informative enough to adapt multiple hypothesis testing. If the genes were stochastically independent one could have used the formula

$$P_0\{T_n^* \leq c\} = [P_0\{T_{n1} \leq c\}]^K, \tag{12}$$

so that the distribution-free nature of the T_{nk} under the null hypothesis provides the access to the computation of the test function and the critical level. If n is at least moderately large, in view of the asymptotic normality of T_{nk}, the randomization test function may be replaced by a conventional normal theory test function, where for the individual tests, a significance level α^* is so chosen that $\alpha = 1 - (1 - \alpha^*)^K$. Generally, if we let $\alpha^* = (\alpha/K)$, then the size of the UIT is $\leq \alpha$ no matter whether the T_{nk} are stochastically independent or not. There is, therefore, certain amount of conservativeness in this specification. In passing, we may remark that by the classical asymptotics on Hoeffding's U-statistics, for any pair $(k \neq q)$, (T_{nk}, T_{nq}) being a bivariate U-statistic, for α^* sufficiently small, using the bivariate extreme statistics results (viz., Sibuya 1959), we can claim that the events $\{T_{nk} > c_{\alpha^*}\}$ and $\{T_{nq} > c_{\alpha^*}\}$ will be asymptotically (as $K \to \infty$) independent so that $P_0\{T_{nk} > c_{\alpha^*}, T_{nq} > c_{\alpha^*}\}$ can be well approximated by $[P_0\{T_{nk} > c_{\alpha^*}\}]^2$. This quasi-independence at the high level of crossing may be used to prescribe the following testing procedure:

> *For a chosen $\alpha^* = K^{-1}\alpha$, obtain the marginal distributional critical level c_{α^*}, and reject those H_{0k}; $k \in \{1,\ldots,K\}$ for which the corresponding T_{nk} exceeds c_{α^*}.*

A randomization test function can be prescribed when n is not adequately large. Thus, the UIT provides a bound on the FWER. If $P_1, \ldots, P_K$ are the p-values for the K marginal tests and $P_{K:1} \leq \cdots \leq P_{K:K}$ be the corresponding order statistics, then assuming that under H_0 the P_k have uniform (0, 1) distribution (i.e., tacitly assuming that the T_{nk} have continuous distribution under H_0), Simes'(1986) Theorem, an equivalent version of the classical Ballot theorem (Karlin 1969), asserts that for every $\alpha : 0 < \alpha < 1$,

$$P\{P_{K:k} > k\alpha/K, \ \forall \ k = 1, \ldots, K \ |H_0\} = 1 - \alpha. \tag{13}$$

Suppose now we define the *anti-ranks* $S_1, \ldots, S_K$ by letting

$$P_{K:k} = P_{S_k}, \ k = 1, \ldots, K, \tag{14}$$

where again ties among the ranks are neglected under the assumption of continuity of the distribution of the P_k. Whereas Simes' theorem provides a test of the overall hypothesis, Hochberg (1988) derived a step-up procedure for multiple hypotheses testing based on the following
For every $\alpha \in (0, 1)$,

$$P\{P_{K:k} \geq \alpha/(K - k + 1), \ \forall \ k = 1, \ldots, K \ |H_0\} = 1 - \alpha. \tag{15}$$

Benjamini and Hochberg (1995) considered a step-up procedure based on the Simes theorem. Their multiple hypothesis testing procedure is the following:

> *Reject those null hypotheses $\{H_{0S_k}\}$ for which $P_{S_k} \leq k\alpha/K$, $k = 1, \ldots, K$, and accept the other null hypotheses in the complementary set.*

For some related developments in a parametric setup, we refer to Benjamini and Hochberg (1995), Dudoit et al. (2003), Lehmann and Romano (2005), Storey (2007) and Sarkar (2006), among others. These developments paved the way for other measures of error rates (including the commonly used *False discovery rate* (FDR)) which are more adaptable in the $K >> n$

environment. For details, we may refer to Sen (2008). These developments rest on some assumptions that are not likely to be tenable when n is small and when the test statistics have discrete distributions. As such, we take recourse to some modification of the Chen-Stein theorem (viz., Sen 2008) which covers discrete distributions and possibly small sample size too.

Chen-Stein Theorem: Let $\mathcal{I}$ be an index set with elements $i \in \mathcal{I}$ and let K be the cardinality of the set $\mathcal{I}$. For each $i \in \mathcal{I}$ let Y_i be an indicator random variable and let

$$P\{Y_i = 1\} = 1 - P\{Y_i = 0\} = p_{Ki}, \ i \in \mathcal{I}. \tag{16}$$

Let $W = \sum_{i \in \mathcal{I}} Y_i$ the total number of occurrence of the events $\{Y_i = 1\}$, $i \in \mathcal{I}$, and let $\lambda_K = \sum_{i \in \mathcal{I}} p_{Ki} = E(W)$. For each $i \in \mathcal{I}$, we define a set $\mathcal{J}_i \in \mathcal{I}$ and its complement $\mathcal{J}_i^c$ as the set of dependence of i and its complement, set of independence of i. Thus, it is tacitly assumed that Y_i is independent of $\{Y_j, \ j \in \mathcal{J}_i^c\}$, for every $i \in \mathcal{I}$. Further, let

$$\begin{aligned} b_1 &= \sum_{i \in \mathcal{I}} \sum_{j \in \mathcal{J}_i} E(Y_i)E(Y_j); \\ &= \sum_{i \in \mathcal{I}} \sum_{j \in \mathcal{J}_i} p_{Ki} p_{Kj}, \end{aligned} \tag{17}$$

$$b_2 = \sum_{i \in \mathcal{I}} \sum_{j(\neq i) \in \mathcal{J}_i} E(Y_i Y_j), \tag{18}$$

and

$$b_3 = \sum_{i \in I} E|\{E(Y_i - E(Y_i)|\{Y_j, \forall j \in \mathcal{J}_i^c\})|. \tag{19}$$

Finally, let Z be a random variable having Poisson distribution with parameter $E(Z) = \lambda_K$. Then

$$\begin{aligned} &||\mathcal{L}(W) - \mathcal{L}(Z)|| \\ &\quad \le 2(b_1 + b_2 + b_3)\frac{1 - e^{-\lambda_K}}{\lambda_K} \\ &\quad \le 2(b_1 + b_2 + b_3)\min\{1, \lambda_K^{-1}\}. \end{aligned} \tag{20}$$

A direct corollary to the Theorem is the following:

$$|P\{W = 0\} - e^{-\lambda_K}| \le 2(b_1 + b_2 + b_3)\min\{1, \lambda_K^{-1}\}. \tag{21}$$

An interesting feature of this theorem is the dual control of λ_K, the expectation and b_1, b_2, b_3 the dependence functions. In line with our intended application of this basic result, consider a natural extension of this result

(Sen 2008) covering both continuous and discrete time parameter Poisson process approximations. In the present context, under the null hypothesis, all the T_{nk} have a common distribution, discrete but symmetric about 0, and is completely known (though could be computationally intensive if n is not too small). Let us denote the distinct mass points for T_{nk} by $-1 = a_1 < a_2 < \cdots, a_L = 1$ and let

$$\tau_j = P_0\{T_{nk} \geq a_{L-j+1}\},\ j = 1, \ldots, L. \tag{22}$$

Then $0 \leq \tau_1 < \tau_2 < \cdots < \tau_L \leq 1$. Also, let us write

$$Y_k(\tau_j) = I(T_{nk} \geq a_{L-j+1}),\ j = 1, \ldots, L,\ k = 1, \ldots, K. \tag{23}$$

Further, let

$$W_K(\tau_j) = \sum_{k=1}^{K} Y_k(\tau_j),\ j = 1, \ldots, L. \tag{24}$$

Also, let $J = \max\{j : 1 \leq j \leq L;\ \tau_j \leq \eta\}$ for some prefixed $\eta > 0$. Basically, we would like to pursue the distributional features of the partial sequence: $\{W_K(\tau_j),\ j \leq J\}$. Corresponding to the known points $\tau_1 < \cdots < \tau_J$, let us consider the partial process $W_K(\tau_j), j = 1, \ldots, J$ as defined above. Also, let us choose a set of nonnegative integers $r_1 \leq \cdots \leq r_J$ in such a way that

$$P_0\{W_K(\tau_j) > r_j, \text{for some } j \leq J\} = \alpha, \tag{25}$$

where α may not be exactly equal to a specified level (such as 0.05) but can be approximated very well through the above Poisson process result. This allows a multi-stage multiple hypothesis testing procedure, illustrated in detail in Sen (2008). We may then consider the following testing procedure.

> *Compute the $W_K(\tau_j),\ j \leq J$ as above. If $W_K(\tau_j) \leq m_j,\ \forall j \leq J$, accept the null hypothesis that there is no DG. On the other hand, if $W_K(\tau_j)$ is greater than m_j for at least one $j \leq J$, then reject the null hypothesis that all the genes are NDG, and proceed to detect those genes $k \in \mathcal{K}$as DG where*

$$\mathcal{K} = \{k \in \{1, \ldots, K\} : Y_k(\tau_j) = 1,\ \text{for some } j \leq J\}. \tag{26}$$

Note that if for some $k(= 1, \ldots, K)$, $Y_k(\tau_j) = 1$ for some $j \leq J$, then $Y_k(\tau_{j'}) = 1,\ \forall j' \geq j$. Further, note that $\mathcal{K}$ is a stochastic subset of $\{1, \ldots, K\}$, and $R =$ cardinality of $\mathcal{K}$ is a (nongenative) integer valued random variable. The overall significance level of this testing procedure is well approximated by the preassigned level α. Further, though based on numerical studies, this multi-stage testing procedure seems to have a better FDR and other properties (Kang and Sen 2008). The main advantage of this approach is that no homogeneity across the genes is needed to use such coordinatewise distribution-free statistics, and much less restrictive dependence conditions are needed in this approach. For qualitative data models, even coordinatewise statistics may not be genuinely distribution-free under the null hypothesis. However, the permutation distribution of these statistics provides a working formula where the multi-stage version of the Chen-Stein theorem works out neatly.

References

Arratia,R.,Goldstein, L., and Gordon, L. (1990). Poisson approximation and the Chen-Stein method : Rejoinder. *Statistical Science 5*, 432-434.

Benjamini, Y. and Hochberg, Y. (1995). Controlling the false discovery rate: a practical and powerful approach to multiple testing. *Jour. Royal Statist. Soc. B 57*, 289-300.

Chen, L.H.Y. (1975). Poisson approximation for dependent trials. *Ann. Probab. 3*, 534-545.

Dudoit, S., Shaffer, J., and Boldrick, J. (2003). Multiple hypothesis testing in microarray experiments. *Statist. Sci. 18*, 71 - 103.

Hochberg, Y. (1988). A sharper Bonferroni procedure for multiple tests of significance. *Biometrika 75*, 800-802.

Hoeffding, W. (1948). A class of statistics with asymptotically normal distribution. *Ann. Math. Statist. 19*, 293-325.

Kang, M. and Sen, P.K. (2008). Kendall's tau-type rank statistics in genome data. *Appl. Math. 53*, 207-221.

Karlin, S. (1969). *A First Course in Stochastic Processes.* Academic Press, New York.

Lehmann, E.L. and Romano, J.P. (2005). Generalizations of the familywise error rate. *Ann. Statist. 33*, 1138-1154.

Lobenhofer, B.K., Bennett, L., Cable, P.L., Bushel, P.R., and Afshari, C.A. (2002). Regulation of DNA replication fork genes by 17-estradiol. *Molecular Endocrinology 16*, 1215-1229.

Roy, S. N. (1953). A heuristic method of test construction and its use in multivariate analysis. *Ann. Math. Statist. 24*, 220-238.

Sarkar, S.K. (2006). False discovery and false nondiscovery rates in single-step multiple testing procedures. *Ann. Statist. 34*, 394-415.

Sen, P.K. (1968). Estimates of regression coefficients based on Kendall's tau. *Jour. Amer. Statist. Assoc. 63*, 1379-1389.

Sen, P. K. (1981). *Sequential Nonparametrics: Invariance Principles and Statistical Inference.* John Wiley, New York.

Sen, P.K. (2004). *Excursions in Biostochastics: Biometry to Biostatistics to Bioinformatics* Invited Lecture Ser. No.5, Institute of Statistical Studies, Academia Sinica, Taipei.

Sen, P. K. (2006). Robust statistical inference for high-dimensional data models with applications to genomics. *Austrian Jour. Statist. 35*, 197-214.

Sen, P. K. (2008). Kendall's tau in high-dimensional genomic parsimony. *Inst. Math. Statist. Collection, 3*, 251 - 266.

Sen, P. K, Tsai, M.-T. and Jou, Y.-S. (2207). High-dimension low sample size perspectives in constrained statistical inference : The SARSCoV RNA genome in illustration. *Jour. Amer. Statist. Assoc. 102*, 686-694.

Sibuya, M. (1959). Bivariate extreme statistics. *Ann. Institut. Statist. Math. 11*, 195-210.

Silvapulle, M. J. and Sen, P. K. (2004). *Constrained Statistical Inference: Inequality, Ordered, and Shape Restraints.* John Wiley, New York.

Simes, R.J. (1986). An improved Bonferroni procedure for multiple tests of significance. *Biometrika 73*, 751-754.

Storey, J. (2007). The optimal discovery procedure: A new approach to simultaneous significance testing. *Jour. Roy. Statist. Soc. B69*,1 - 22.

STUDIES OF NONLINEAR THREE-DIMENSIONAL FREE SURFACE FLOWS

J.-M. VANDEN-BROECK*

Department of Mathematics, University College London,
London, WC1E 6BT, UK
** E-mail: broeck@math.ucl.ac.uk*

New types of two and three-dimensional free surface flows are presented and discussed. The solutions are obtained by boundary integral equation methods. The discretized equations are solved by Newton's iterations coupled with continuation methods. Appropriate initial guesses for the iterations are found by first solving a 'modified problem' and then taking an appropriate limit.

Keywords: free surface flows, water waves, three-dimensional solitary waves.

1. Introduction

Periodic two-dimensional waves propagating at a constant velocity at the surface of a fluid bounded below by a horizontal bottom have been studied for more then 150 years. Here the waves have wavelength λ and extend from $x = -\infty$ and $x = \infty$ (see Fig. 1). The top curve represents the free surface (i.e. the interface between the fluid and the atmosphere, which is characterized by a constant pressure). The fluid is usually assumed to be incompressible and inviscid and the flow to be irrotational.

A motivation for this problem is the two-dimensional flow generated by an object moving at a constant velocity below a free surface. An example is shown in Fig. 2. Here the object is a half-cylinder of circular cross section at the bottom of the channel. The cylinder is assumed to be very long and Fig. 2 represents a cross section far away from the ends of the cylinder. The flow is then approximately two-dimensional. For appropriate choice of the parameters, waves appear on the free surface. These waves approach in the far field trains of waves of constant amplitudes similar to the one sketched in Fig. 1. In the example of Fig. 2 gravity is taken into account and surface tension is neglected. The free surface is then flat as $x \to -\infty$ and there is a train of waves of constant amplitude as $x \to \infty$.

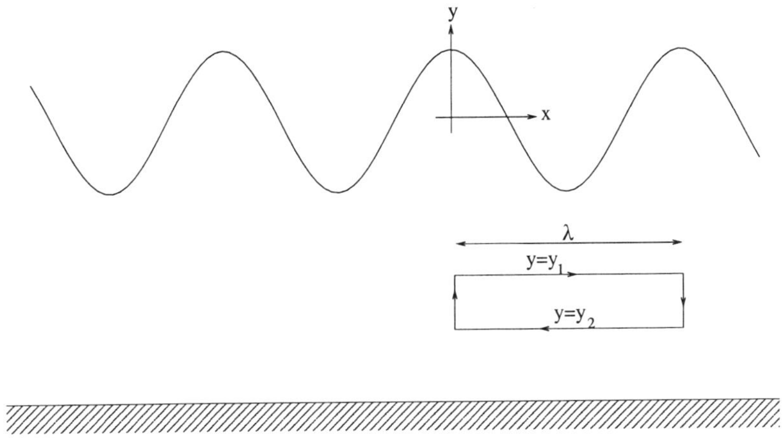

Fig. 1. Sketch of a two-dimensional train of waves viewed in a frame of reference moving with the wave. The free surface profile has wavelength λ. The fluid is bounded below by a horizontal bottom.

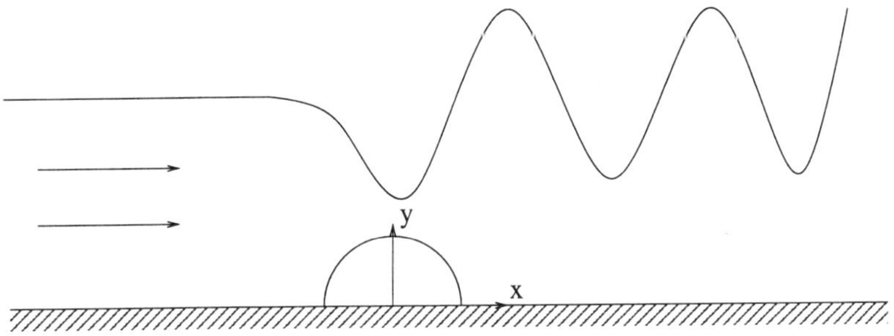

Fig. 2. Sketch of the two-dimensional free surface flow past a submerged circle.

As the wavelength λ becomes very large compared to the depth of the fluid, the waves of Fig. 1 approach solitary waves. When surface tension is neglected, these solitary waves look like simple 'humps' and are characterized by a flat free surface in the far field (see Fig. 3). They exist as solutions of the full Euler equations and are described by the classical sech^2 solution of the Korteweg de Vries equation when their amplitudes is small.

In this paper we present various extensions of the solitary wave sketched in Fig. 3. These include the effects of surface tension, three dimensional calculations and the study of the influence of disturbances. One basic method used in the computations is to first calculate solutions of a 'modified problem' and then to obtain the solution of the original problem by taking

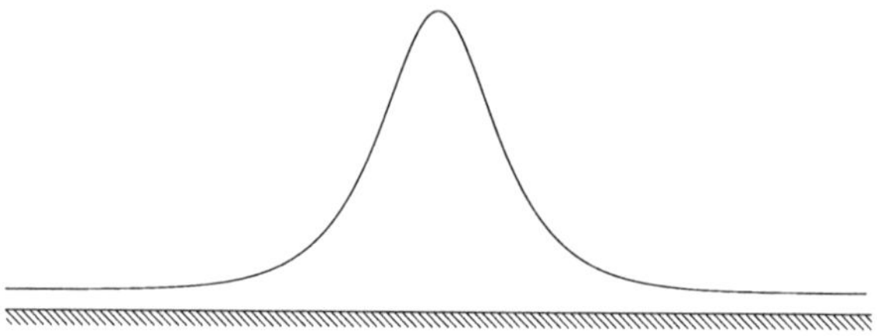

Fig. 3. Sketch of a solitary wave.

numerically an appropriate limit.

2. Two-dimensional gravity-capillary solitary waves

When surface tension is included in the dynamic boundary condition, there are no solitary waves similar to that of Fig. 3: all the solitary waves are characterized by oscillations in the far field. These oscillations can be of constant amplitudes (see Figure 4) or of decaying amplitudes (see Figs. 5 and 6).

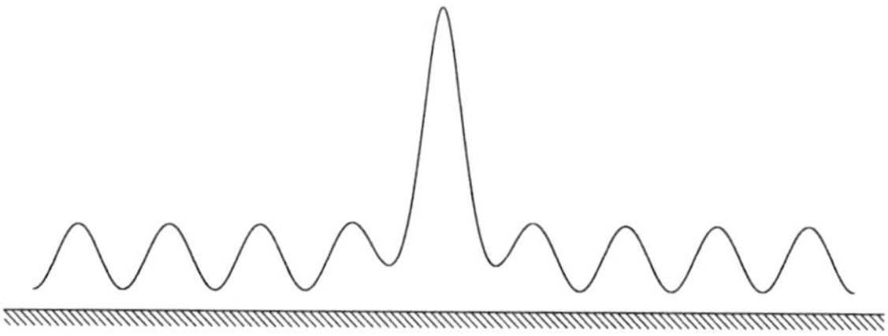

Fig. 4. Sketch of a generalized solitary wave.

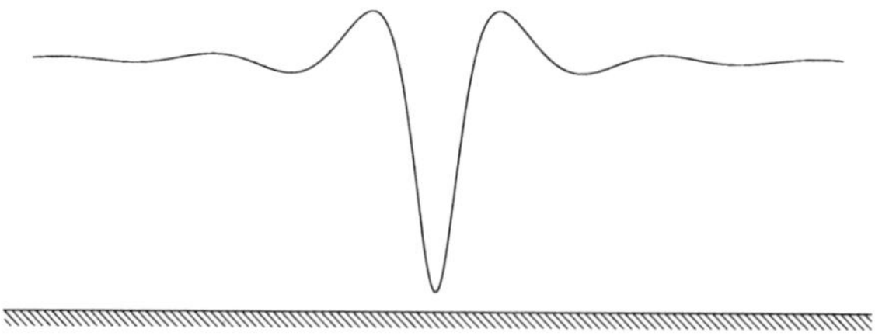

Fig. 5. Sketch of a depression solitary wave with decaying tail.

When the amplitude of the oscillation is constant (Fig. 4), the solitary waves are referred to as generalized solitary waves to distinguish them

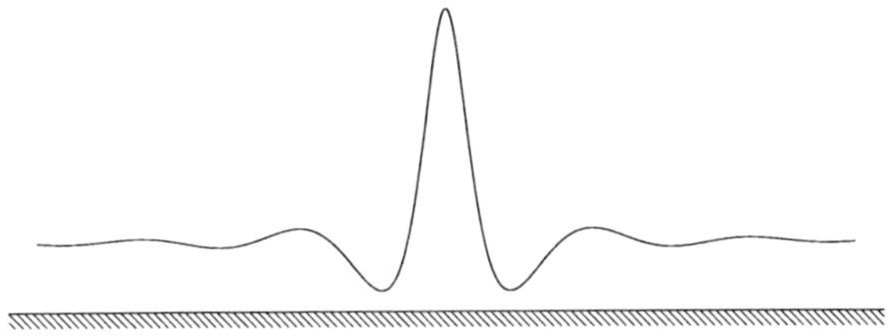

Fig. 6. Sketch of an elevation solitary wave with decaying tail.

from true solitary waves which are flat in the far field. Generalized solitary waves have been calculated numerically by Hunter and Vanden-Broeck[1] and Champneys *et al.*[2] There are also many theoretical results (see Dias and Khariff[3] for a review). One interesting mathematical feature is that the amplitude of the oscillations become exponentially small as the amplitude of the solitary wave tends to zero. We shall not consider further these generalized solitary waves.

When the amplitude of the oscillations is decaying, there are both depression waves (see Fig. 5) and elevation waves (see Fig. 6). In this section we show how to compute solutions corresponding to the sketches. Then in Section 3 we shall generalized the method to show numerically the existence of three dimensional gravity capillary solitary waves.

2.1. *Formulation*

We assume the fluid to be incompressible and inviscid and the flow to be irrotational. For simplicity we also assume that the depth of the fluid is infinite (extensions to finite depth will be mentioned later). We take a frame of reference moving with the waves of Fig. 5 and Fig. 6 so that the flow is steady. The velocity can then be written as the gradient of a potential function ϕ satisfying Laplace equation

$$\phi_{xx} + \phi_{yy} = 0 \tag{1}$$

in the flow domain with the boundary conditions

$$\phi_y = \phi_x \eta_x \quad \text{on} \quad y = \eta(x) \tag{2}$$

$$\frac{1}{2}(\phi_x^2 + \phi_y^2) + g\eta(x) - \frac{T}{\rho}\frac{\eta_{xx}}{(1+\eta_x^2)^{3/2}} = B \quad \text{on} \quad y = \eta(x) \tag{3}$$

$$\phi_x \to U \quad \text{as} \quad y \to -\infty \tag{4}$$

Here T is the surface tension, g the acceleration of gravity, ρ the density of the fluid, $y = \eta(x)$ the (unknown) equation of the free surface and B the Bernoulli constant. Equations (2) and (3) are the kinematic and dynamic boundary conditions on the free surface. We chose the origin of y as the free surface at infinity.

2.2. *The modified problem and the numerical solutions*

The problem defined by Eqs. (1)-(4) can be reformulated as an integro-differential equation involving only unknowns on the free surface (see Vanden-Broeck and Dias[4] and Dias *et al.*[5]). This equation can then be discretized and solved numerically by Newton's iterations. The main difficulty in computing the gravity-capillary solitary waves of Figs. 5 and 6 is to find an appropriate initial guess. To do this we follow Vanden-Broeck and Dias[4] and define a modified problem by replacing Eq. (3) by

$$\frac{1}{2}(\phi_x^2 + \phi_y^2) + g\eta(x) - \frac{T}{\rho}\frac{\eta_{xx}}{(1+\eta_x^2)^{3/2}} + \epsilon P(x) = B \quad \text{on} \quad y = \eta(x) \tag{5}$$

The term $\epsilon P(x)$ in Eq. (5) represents a prescribed distribution of pressure on the free surface. It can be viewed as an inverse method to compute the free surface flow past an obstacle (e.g. a ship) moving at a constant velocity U at the surface of a fluid since the portion of the free surface under the support of the distribution of pressure can be replaced by a rigid surface. However our main concern here is to solve the problem (1), (2), (5) and (4) and to take numerically the limit $\epsilon \to 0$.

We present in Fig. 7 numerical values of

$$a = \frac{g\eta_M}{U^2} \tag{6}$$

versus

$$\alpha = \frac{Tg}{\rho U^4} \tag{7}$$

Here η_M corresponds to the largest value $|\eta|$ on the free surface. The top curve corresponds to a small negative value of ϵ and the bottom curve to a small positive value of ϵ. Both curves have a turning point. We denote the values of α corresponding to the turning points by α_{up} and α_{down} where *up* and *down* refer to the upper and lower curve of Fig. 7.

We recall that linear gravity capillary waves in infinite depth are characterized by the dispersion relation

$$U^2 = \frac{g}{k} + \frac{T}{\rho}k \tag{8}$$

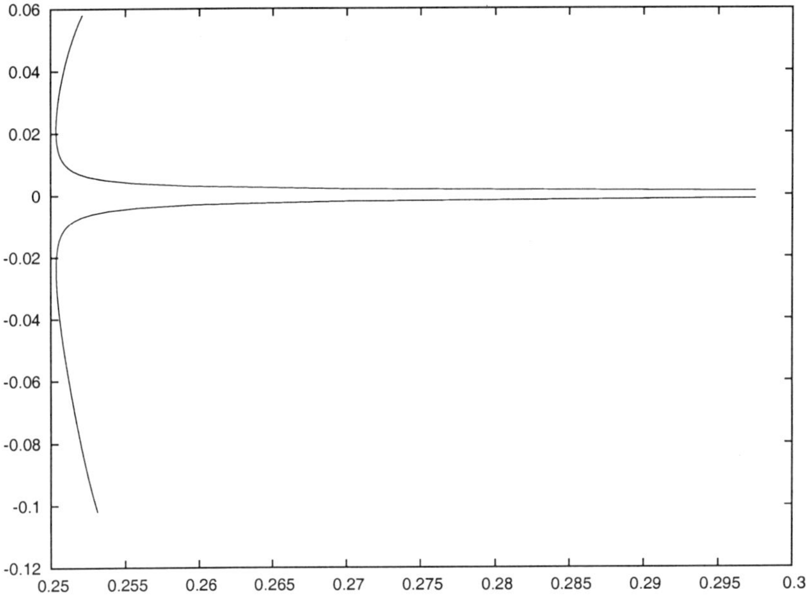

Fig. 7. Values of a versus α.

where k is the wave number. It can easily be checked that Eq. (8) has only real roots for k provided $U > c_{min}$ where

$$c_{min} = 2(\frac{Tg}{\rho})^{1/2} \tag{9}$$

Using Eq. (7) and Eq. (9) we see that

$$\alpha < \frac{1}{4} \quad \text{when} \quad c < c_{min} \tag{10}$$

and

$$\alpha > \frac{1}{4} \quad \text{when} \quad c > c_{min} \tag{11}$$

Therefore both curves in Fig. 7 are in the region $\alpha > 1/4$ where there are no linear gravity capillary waves (see (the inequality (11))).

The solutions corresponding to the portions ($\alpha > \alpha_{down}$, $\alpha > \alpha_{up}$) of the curves of Fig. 7 closer to the α-axis are perturbations of a uniform stream. This can be shown in the following way. Select a solution corresponding to a point on the portions of the curves of Fig. 7 (let us say for $\alpha = 0.28$). Then use this solution as an initial guess to compute a new solution for $\alpha = 0.28$ and a value ϵ_1 of ϵ slightly smaller than the one in Fig. 7. Use then this solution to compute a new solution for a value of ϵ slightly smaller than ϵ_1 and so on. Ultimately we obtain a solution corresponding to $\epsilon = 0$. This solution is a uniform stream.

We refer to this procedure of computing solutions for a decreasing sequence of values of ϵ as a continuation in ϵ (or in which ever parameter we use).

An interesting question is what happens if we use the continuation in ϵ by starting with a solution on the portions of the branches of Fig. 7 (for $\alpha > \alpha_{up}$ and $\alpha > \alpha_{down}$) which are further from the α-axis. The answer is that the resulting solutions for $\epsilon = 0$ are the solitary waves sketched in Figs. 5 and 6. These solitary waves form two branches of solutions bifurcating from the α-axis at $\alpha = 1/4$ in Fig. 7. The first branch is in the upper half plane of Fig. 7 and corresponds to the elevation solitary wave sketched in Fig. 6. The second branch is in the lower half plane of Fig. 7 and corresponds to the depression solitary wave sketched in Fig. 5. These branches are computed by continuation once a particular solution has been obtained by continuation in ϵ. The branches are not shown in Fig. 7 (details can be found in Vanden-Broeck and Dias[4] and Dias *et al.*[5]). However the corresponding branches for three-dimensional waves will be presented in Section 3.

Our findings can be summarized as follows. For $\epsilon = 0$, the solutions in Fig. 7 consist of the α-axis (uniform stream) and of the branches of solitary waves bifurcating from the α-axis at $\alpha = 1/4$ and extending for $\alpha > 1/4$. For $\epsilon \neq 0$, the curves in Fig. 7 have turning points at $\alpha = \alpha_{up}$ and $\alpha = \alpha_{down}$. (the values of α_{up} and α_{down} depend on ϵ). The solutions corresponding to the portions of the curves closer to the α-axis are perturbations of a uniform stream, whereas the solutions corresponding to the portions of the curves further from the α-axis are perturbations of the solitary waves. Typical free surface profiles of solitary waves are shown in Figs. 8 and 9.

3. Three-dimensional gravity-capillary solitary waves

The numerical procedure described in Section 2.2 for two-dimensional solitary waves can be generalized for three-dimensional waves. A modified problem is again defined by introducing a pressure term ϵP in the dy-

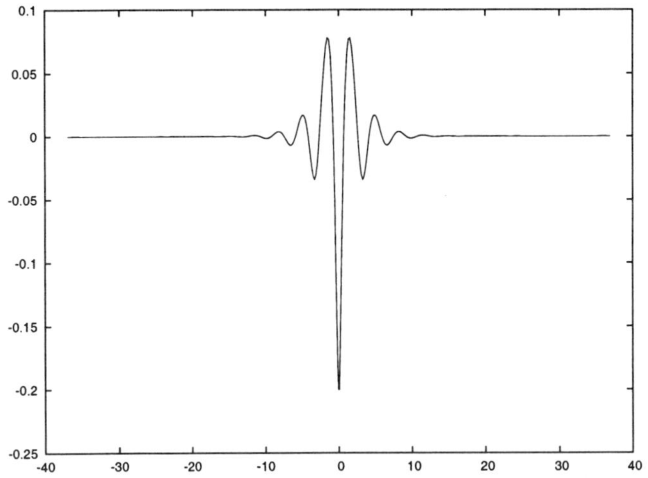

Fig. 8. Free surface profile of a two-dimensional depression solitary wave for $\alpha = 0.264$.

namic boundary condition and then taking numerically the limit $\epsilon \to 0$. The equations are solved by a boundary integral equation method based on free spaced Green's functions (see Forbes[6] and Parau, Vanden-Broeck and Cooker[7–9]).

As for the two-dimensional problem of Section 2.2, there are two branches of solitary waves bifurcating from $\alpha = 1/4$. They are shown in Fig. 10 where we plot values of

$$\tilde{a} = \frac{\rho U^2}{T}\eta_M \tag{12}$$

versus α. The parameter $\tilde{a}$ is related to the parameter a defined in Eq. (6) by $a = \tilde{a}\alpha$. The first branch is in the upper plane and corresponds to elevation solitary waves. The second branch is in the lower plane and corresponds to depression solitary waves.

A typical profile of a depression solitary wave is shown in Fig. 11.

The wave propagates in the x-direction. The profile has decaying oscillations in the direction of propagation and monotonic decay in the direction perpendicular to the direction of propagation. These properties are clearly shown in Fig. 12 where we present half of the wave profile.

A typical profile for an elevation solitary wave is shown in Fig. 13. Again

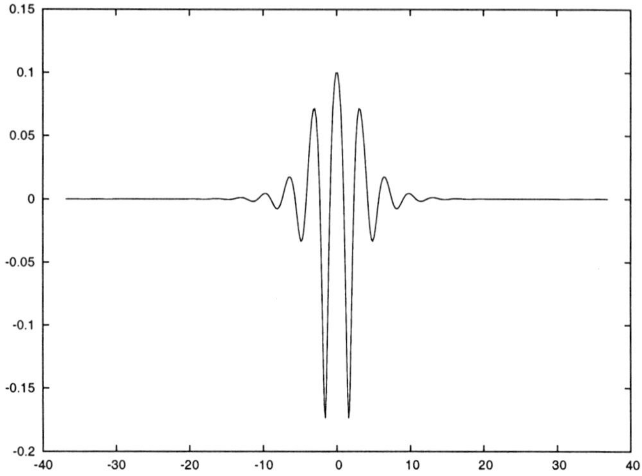

Fig. 9. Free surface profile of a two-dimensional elevation solitary wave for $\alpha = 0.261$.

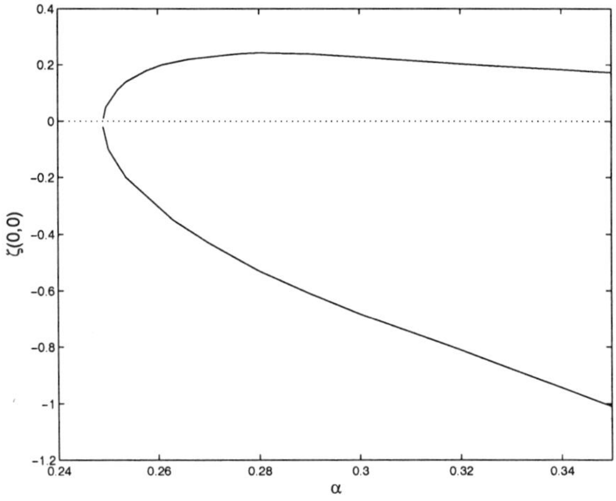

Fig. 10. Values of $\tilde{a}$ versus α.

only half of the wave is shown.

The results presented in this section show that there are three-dimensional gravity capillary solitary waves in water of infinite depth whose

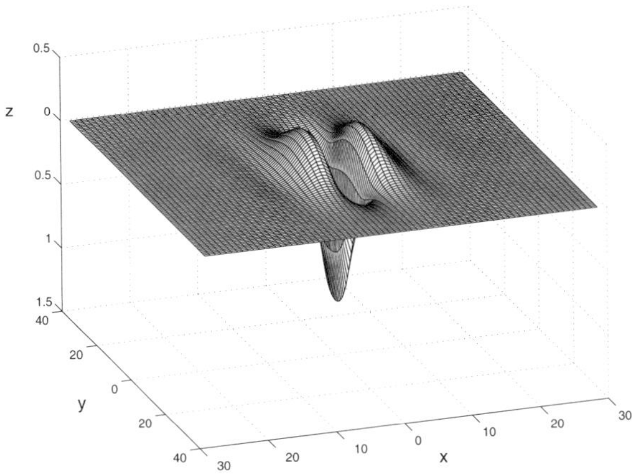

Fig. 11. Free surface profile of a three-dimensional depression solitary wave for $\alpha = 0.35$.

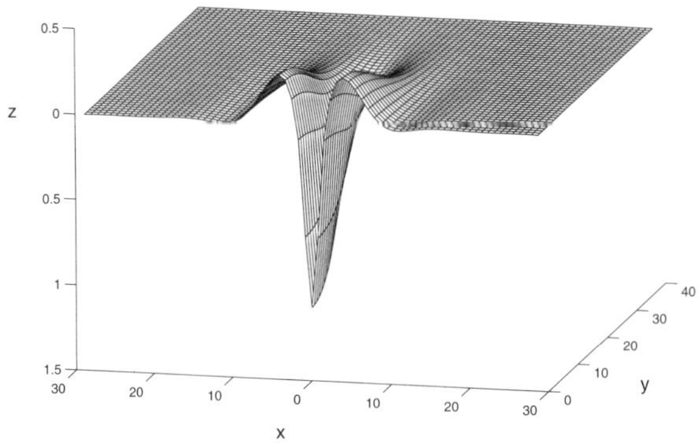

Fig. 12. Free surface profile of a three-dimensional depression solitary wave for $\alpha = 0.35$. Only half of the wave is shown

properties are similar to the two dimensional waves of Section 2.2. These numerical findings were extended to water of finite depth and to interfacial waves in Parau, Vanden-Broeck and Cooker.[10,11]

Let us conclude this section by mentioning that our numerical findings are consistent with rigorous analytical results of Groves and Sun[13] and with the asymptotic calculations of Kim and Akylas[14,15] and Milewski.[16]

4. Further applications of the 'modified problem' method

One of the basic ideas in the computations of Sections 2.2 and 3 was to define a modified problem involving a parameter (denoted by ϵ). Solutions

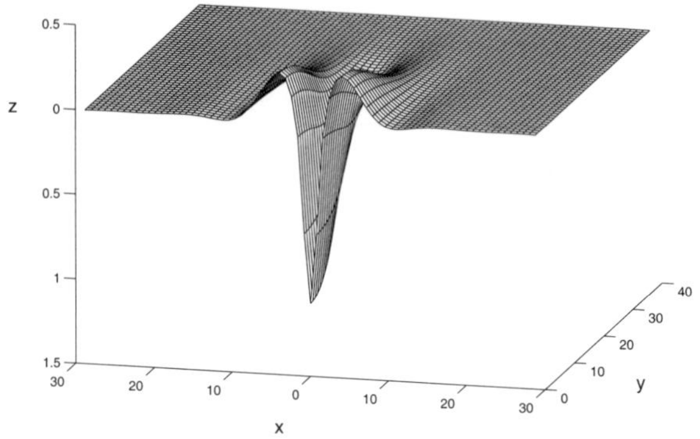

Fig. 13. Free surface profile of a three-dimensional depression solitary wave for $\alpha = 0.345$. Only half of the wave is shown.

were first calculated with $\epsilon \neq 0$. Then the solutions of the original problem were found by taking numerically the limit $\epsilon \to 0$. This method has been used successfully by the author to investigate other free boundary problems.

A first example is the computation of three-dimensional free surface flows generated by moving disturbances in the region $\alpha < 1/4$. There the radiation condition was imposed by first solving a problem with a small amount of dissipation and then taking numerically the limit as the dissipation tends to zero. (see Parau, Vanden-Broeck and Cooker[11]).

A second example is the selection of solutions by taking the limit as the surface tension T tends to zero. It was found that there are problems with have a continuum of solutions when $T = 0$ but a discrete set of solutions when $T \neq 0$. These include fingering in a Hele Shaw cell[17] and models for rising bubbles.[18–23] The modified problems are defined by allowing the finger or bubbles to be pointed. This generates a continuous curve in parameter space. The solutions of the original problem (i.e. the discrete set of solutions) is then found by identifying along the curves the solutions for which the discontinuity in slope vanishes. The solutions of the original problem (i.e. the discrete set of solutions) is then found by identifying along the curve the solutions for which the discontinuity in slope vanishes.

A third example is the computation of water waves in the presence of constant vorticity. By defining appropriate modified problems, new branches of solutions were found (see Vanden-Broeck[24–26]).

References

1. Hunter, J.K. & Vanden-Broeck, J.-M. 1983a Solitary and periodic

gravity-capillary waves of finite amplitude *J. Fluid Mech.* **134**, 205-219.
2. CHAMPNEYS, A.R., VANDEN-BROECK, J.-M. & LORD, G.J. 2002 Do true elevation gravity-capillary solitary waves exist? A numerical investigation *J. Fluid Mech.* **454**, 403-417.
3. DIAS, F. & KHARIF, C. 1999 Nonlinear gravity and capillary–gravity wav es. *Ann. Rev. Fluid Mech.* **31** 301–346.
4. VANDEN-BROECK, J.-M. & DIAS, F. 1992 Gravity-capillary solitary waves in water of infinite depth and related free-surface flows. *J. Fluid Mech.* **240**, 549 - 557.
5. DIAS, F., MENASCE, D. & VANDEN-BROECK, J.-M. 1996 Numerical study of capillary-gravity solitary waves. *Eur. J. Mech., B/Fluids* **15**, 17-36.
6. FORBES, L.K 1989 An algorithm for 3-dimensional free-surface problems in hydrodynamics. *Journal of Computational Physics* **82**, 330-347.
7. PARAU, E. & VANDEN-BROECK, J.-M. 2002 Nonlinear two- and three-dimensional free surface flows due to moving disturbances. *Eur. J. Mechanics B/Fluids*, **21**, 643-656.
8. PARAU, E., VANDEN-BROECK, J.-M. & COOKER, M. 2005a Nonlinear three dimensional gravity capillary solitary waves *J. Fluid Mech.* **536**, 99-105.
9. PARAU, E., VANDEN-BROECK, J.-M. & COOKER, M. 2005b Three-dimensional gravity-capillary solitary waves in water of finite depth and related problems *Phys. Fluids* **17**, 122101.
10. PARAU, E., VANDEN-BROECK, J.-M. & COOKER, M. 2007 Three dimensional gravity and gravity-capillary interfaical flows *Mathematics and Computers in Simulation* **74**, 105-122.
11. PARAU, E., VANDEN-BROECK, J.-M. & COOKER, M. 2007 Nonlinear three-dimensional interfacial flows with a free-surface *J. Fluid Mech.* **91**, 481-494.
12. PARAU, E., VANDEN-BROECK, J.-M. & COOKER, M. 2007 Three-dimensional capillary-gravity waves generated by a moving disturbance *Phys. Fluids* **19**.
13. GROVES, M.D. & SUN, M.S. 2006 Fully localised solitary-wave solutions of the three-dimensional gravity-capillary water-wave problem. *Arch. Rat. Mech. Anal. (submitted)*
14. KIM. B. & AKYLAS, T.R. 2005 On gravity-capillary lumps. *J. Fluid Mech.* **540**, 337–351.
15. KIM. B. & AKYLAS, T.R. 2006 On gravity-capillary lumps Part2 Two dimensional Benjamin equation. *J. Fluid Mech.* **557**, 237–256.
16. MILEWSKI, P.A. 2005 Three-dimensional localized solitary gravity-capillary waves. *Comm. Math. Sc.* **3**, 89–99.
17. VANDEN-BROECK, J.-M. 1983 Fingers in a Hele-Shaw Cell with surface tension *Phys. Fluids* **26**, 2033-2034.
18. VANDEN-BROECK, J.-M. 1984 Bubbles rising in a tube and jets falling from a nozzle *Phys. Fluids* **27**, 1090-1093.
19. VANDEN-BROECK, J.-M. 1984 Rising bubbles in a two-dimensional tube with surface tension *Phys. Fluids* **27**, 2604-2607.
20. VANDEN-BROECK, J.-M. 1986 Pointed bubbles rising in a two dimensional tube *Phys. Fluids* **29**, 1343-1344.

21. VANDEN-BROECK, J.-M. 1986 A free streamline model for a rising bubble *Phys. Fluids* **29**, 2798-2801.
22. VANDEN-BROECK, J.-M. 1988 Joukovskii's model for a rising bubble *Phys, Fluids* **31**, 974-977.
23. LEE, J.W. AND VANDEN-BROECK, J.-M. 1998 Bubbles rising in an inclined two-dimensional tube and jets falling from along a wall *J. Austral. Math. Soc. B* **39**, 332-349.
24. VANDEN-BROECK, J.-M. 1994 Steep solitary waves in water of finite depth with constant vorticity *J. Fluid Mech.* **274**, 339-348.
25. VANDEN-BROECK, J.-M. 1995 New families of steep solitary waves in water of finite depth with constant vorticity *Eur. J. Mech. B/fluids* **14**. 761-774.
26. VANDEN-BROECK, J.-M. 1996 Periodic waves with constant vorticity in water of infinite depth *IMA J. Appl. Math.* **56**, 207-217.

PART B

Invited Papers

BURSTING IN PITUITARY CELLS

R. BERTRAM

Department of Mathematics and
Programs in Neuroscience and Molecular Biophysics,
Florida State University,
Tallahassee, Florida 32306, USA
E-mail: bertram@math.fsu.edu
www.math.fsu.edu

Bursting electrical oscillations are ubiquitous in nerve cells, and have been the focus of mathematical modeling and analysis for three decades. A key feature of these oscillations is the separation of time scales between "fast" and "slow". This separation allows one to perform a geometric singular perturbation or "fast/slow" analysis on the system. This analysis has led to many valuable insights into the dynamic mechanisms of bursting oscillations. Bursting also occurs in hormone-secreting cells of the pituitary. In most cases, however, the oscillation differs significantly from what is observed in nerve cells. In mathematical models of bursting in pituitary cells this difference is due to two factors. First, the underlying bifurcation structure of the fast subsystem is unlike that of neural bursting models. Second, the slow variable(s) is only marginally slow. As a result of these differences, some of the intuition that aids in the understanding of neural bursting can be misleading in the context of pituitary bursting. In this article I highlight some of the differences between the two classes of bursting.

Keywords: bursting; pituitary; neuron; bifurcation.

1. Introduction

Electrical bursting oscillations are characterized by episodes of action potentials or spikes separated by periods of quiescence. They have been described in many neurons, including ganglion neurons of the Aplysia,[1,2] neurons in the pre-Bötzinger complex of the medulla of the brainstem,[3] vasopressin neurons of the hypothalamus,[4] hippocampal pyramidal neurons,[5] and thalamic neurons.[6] Many mathematical models have been developed for bursting neurons, and analysis of the models has been greatly facilitated by the geometric singular perturbation or "fast/slow" analysis technique developed by John Rinzel in the mid 1980s.[7] Here, fast and slow variables are formally separated into fast and slow subsystems. One or more of the slow variables is then used as a bifurcation parameter in the analysis of the bifurcation structure of the fast subsystem. Use of this technique has led to many insights, and various types of bursting oscillations have been classified according to the fast subsystem bifurcation structure.[8–10]

Endocrine cells of the anterior pituitary gland secrete hormones that have many effects on behavior. These cells are electrically excitable, just as neurons are. They are morphologically simpler, since they are roughly spherical cells without the processes that characterize neurons. They also produce electrical bursting oscillations either spontaneously or when a stimulating/inhibiting factor is added/removed.[11] This bursting pattern is, however, quite different from that typically observed in nerve cells. Indeed, the existing neural models and fast subsystem bifurcation structures are ill-suited for the bursting produced by pituitary lactotrophs, somatotrophs, and corticotrophs. Pancreatic β-cells, when isolated from the islets of Langerhans, also exhibit bursting that is similar to that of the pituitary cells.[12]

2. Square Wave Bursting

There are many types of neural bursting oscillations, differing in the bifurcation structures of their fast subsystems. We focus here on type 1 or *square wave bursting*. A minimal model for square wave bursting is:

$$\frac{dV}{dt} = I_{ion}(V, w, c) \tag{1}$$

$$\frac{dw}{dt} = [w_\infty(V) - w]/\tau_w \tag{2}$$

$$\frac{dc}{dt} = \varepsilon g(V) \tag{3}$$

where V is the membrane potential, w is a recovery variable responsible for the downstroke of an action potential, and c is a slowly changing inhibitory variable. In many neural models c represents the free intracellular calcium ion concentration. The function I_{ion} is composed of several terms representing ionic currents. This function, as well as the functions w_∞ and g, is typically nonlinear. The V and w variables make up the *fast subsystem* while the c variable constitutes the *slow subsystem*.

Figure 1 shows a numerical simulation of square wave bursting along with the time course of the slow variable c. During the spiking or active phase c slowly rises until it reaches a sufficiently high level to terminate the spiking. After this the model cell is silent and c slowly declines, until it reaches a sufficiently low level to restart the spiking. A key observation is that for the same value of c the fast subsystem may be either spiking (oscillatory, O) or silent (S). Thus, the fast subsystem is *bistable* with coexisting limit cycle and steady state attractors.

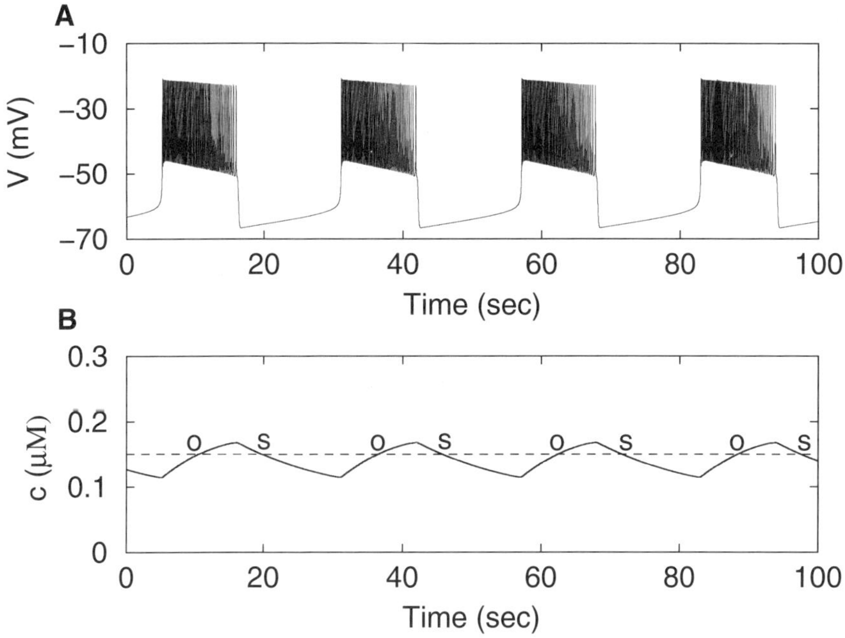

Fig. 1. **(A)** Model simulation of square wave bursting. **(B)** Slow time course of c reveals that the fast subsystem is bistable. The model used is from Ref. 13.

This bursting oscillation can be analyzed by treating c as a parameter of the fast subsystem and computing the fast subsystem bifurcation diagram. Figure 2 shows such a diagram. The z-shaped curve contains stationary solutions; the solid portion of the curve represents stable solutions while the dashed portion represents unstable solutions. Two saddle node (SN) bifurcations occur at turning points. There is also a periodic branch of spiking solutions that terminates at a homoclinic (HM) bifurcation. The two curves representing this branch show the maximum and minimum voltages during the spiking oscillation. The region of bistability between steady state and oscillatory solutions is marked with a double arrow. The next step is to think of the bifurcation diagram as a nullcline in the c-V plane, and also include the c nullcline. Finally, we include the bursting trajectory. Because of the difference in time scales, the trajectory moves along the stable manifolds of the fast subsystem. During the silent phase the phase point travels along the bottom branch of the z-curve until its termination at the SN bifurcation. At this point the trajectory is attracted to the stable periodic branch and now moves rightward (since it is above the c nullcline) until the branch ends at the HM bifurcation. The phase point then returns to the bottom branch and restarts the cycle. This is very similar to a relaxation oscillation, except that now the upper branch of the z-curve is unstable and its role is played by a branch of stable periodic solutions. Two essential ingredients of this square wave bursting are bistability between stationary and spiking solutions in the fast subsystem, and the existence of a variable whose dynamics are much slower than those of the other variables.

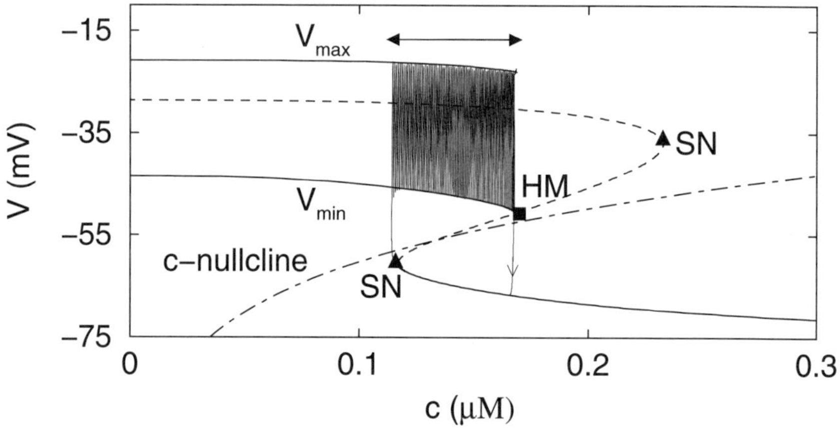

Fig. 2. Fast/slow analysis of square wave bursting. The solid curves are stable stationary or periodic solutions, the dashed curve is unstable stationary solutions, and the dot-dashed curve is the c nullcline.

3. Pituitary Bursting

The right hand sides of Eqs. 1–3 can be adjusted to describe bursting oscillations in pituitary cells. Figure 3A shows a simulation using the model from Ref. 14. (Other pituitary burster models include Refs. 15–17.) In contrast with the square wave bursting, this pituitary-type bursting has only a few spikes and the spikes are small. Also, the burst period is short. The fast subsystem bifurcation diagram is shown in Fig. 3B. As in Fig. 2, the stationary branch of the diagram has two saddle-node bifurcations at turning points. In addition, there is a subcritical Hopf bifurcation (HB) on the top branch of the z-curve. The branch of unstable periodic solutions that emerges from the HB terminates at a homoclinic bifurcation on the middle branch of the z-curve. Beyond the HB, the steady state solutions on the top branch of the z-curve are stable. With this bifurcation structure there are no stable spiking solutions, and the bistability is between two steady state branches. This is one way in which the pituitary burster differs qualitatively from the square wave burster.

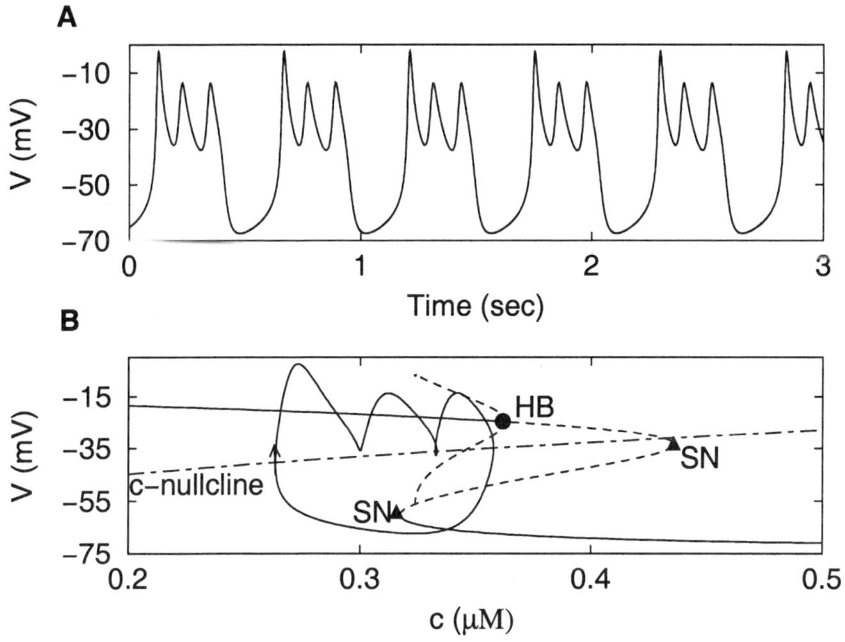

Fig. 3. **(A)** A simulation of pituitary bursting. **(B)** Fast/slow analysis of pituitary bursting. The model used is from Ref. 14 with $g_{BK} = 0.4$ nS.

If c were truly a slow variable, then the oscillation produced would not be bursting, but a relaxation oscillation. The trajectory would travel leftward along a portion of the bottom branch of the z-curve, jump to the top portion of the z-curve at the left SN bifurcation, travel rightward along the top portion of the z-curve, and jump down at the HB bifurcation. However, although c changes more slowly than either V or w in the pituitary model, the difference in time scales is not great. Thus, c could be called a "semi-slow" variable. Because c is only semi-slow, the trajectory is not constrained to travel closely along the z-curve. Instead, it follows the general shape of the curve, while shooting far past the lower SN and re-entering the silent phase before the HB. Most importantly, c changes so rapidly that the phase point does not have time to spiral in to the stable steady states on the top branch. As a result, oscillations are produced during the active phase. These oscillations are the small-amplitude spikes of the pituitary burst. Thus, another key difference between pituitary and square wave bursters is that the square wave burster has a true slow variable, while the equivalent variable in the pituitary burster is only semi-slow.

One result of the differences addressed above is that the resetting properties of square wave and pituitary bursters differ significantly. Because the active and silent phases of square wave bursting each reflect a stable attractor, and because the attractors coexist, it is always possible to reset the oscillation from one phase to the other with a very brief voltage pulse. This manipulation can be performed in the lab, so the resetting properties of the system serve as model predictions that can be tested experimentally. In addition to the prediction that resetting is possible, the model predicts that the phase immediately following the perturbation should be shorter than usual, since the slow variable has not had time to reach its terminal value prior to perturbation. For the pituitary burster, resetting from the active to the silent phase is easy. However, resetting from the silent to the active phase is much more difficult, since the basin of attraction of the upper steady state is small and not always reachable from the lower steady state through a short voltage perturbation of any magnitude. However, because the semi-slow variable changes somewhat rapidly following a voltage perturbation, it is often possible for perturbations with longer duration to reset the system from silent to active. Thus, because c is only semi-slow, the resetting properties of the pituitary burster are more complex than what one would expect from an analysis in which the slow variable is held fixed. This issue is addressed in detail in Ref. 16.

4. Bursting With No Slow Variable

There are many different types of potassium ion channels, giving rise to ionic currents with different properties. One type of K^+ current is the "A-type" K^+ current (I_A). This current activates when the cell membrane is depolarized, but then inactivates quickly. Thus, it is typically activated for only a short time. We have recently shown that this A-type current can serve as the trigger for bursting in a minimal model of the pituitary lactotroph.[17] The model is:

$$\frac{dV}{dt} = I_{ion}(V, w, e) \tag{4}$$

$$\frac{dw}{dt} = [w_\infty(V) - w]/\tau_w \tag{5}$$

$$\frac{de}{dt} = [e_\infty(V) - e]/\tau_e \tag{6}$$

where I_{ion} includes the A-type current, w is a recovery variable as before, and e is an inactivation variable for the A-current. In this model none of the variables can be considered slow or semi-slow.

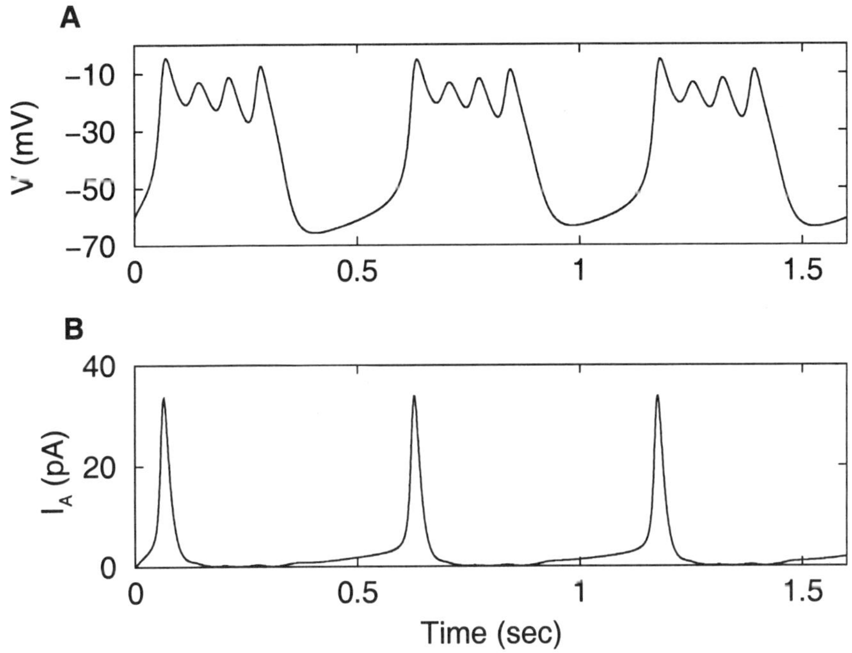

Fig. 4. **(A)** A simulation of pituitary bursting without a slow variable. **(B)** The A-type K^+ current has a surge at the beginning of a burst, which serves as the burst trigger. The model used is from Ref. 17 with $g_A = 13$ nS.

Figure 4A shows bursting produced with the model. This appears to be very similar to the bursting produced by the model with a semi-slow variable (Fig. 3A). However, in this case it is not possible to perform a fast/slow analysis. Instead, one must perform an analysis on the full three-dimensional system.[17] Figure 4B shows the I_A time course. There is a surge of I_A activity at the beginning of each burst, after which the current inactivates for the remainder of the burst. The full I_A surge is necessary for triggering a burst; if the surge is ended prematurely then the number of spikes in the burst is reduced, or the burst is converted to a single spike. The current triggers the burst by injecting the three-dimensional phase point into a neighborhood of a depolarized unstable steady state. The spiraling around this steady state produces the small-amplitude spikes in the burst. Unlike the prior pituitary model, the depolarized state is unstable, so the spiraling is outward. Resetting from the silent to the active phase through a V perturbation is almost impossible with this burst mechanism.

One peculiar property of this type of bursting is that the active phase duration increases when the conductance of the A-current is increased. This is unusual, since in other bursting models increasing a K^+ conductance decreases the active phase duration.

5. Conclusion

While there have been numerous mathematical analyses performed on neural bursting oscillations, there have been only a few mathematical studies of pituitary bursting. These studies have shown that the properties of the two classes of oscillators are quite different, and in some cases entirely different analysis techniques must be employed. Further studies of pituitary bursting should shed light on the mechanisms driving these oscillations, and the properties of the oscillations. They should also provide some interesting mathematical questions, and hopefully some answers.

Acknowledgments

The author's research is funded by NSF grant DMS-0613179 and NIH grant DA-19356.

References

1. P. A. Mathieu and F. A. Roberge, *Can. J. Physiol. Pharmacol.* **49**, 787 (1971).
2. H. M. Pinsker, *J. Neurophysiol.* **40**, 544 (1977).
3. N. Koshiya and J. C. Smith, *Nature* **400**, 360 (1999).
4. P. Roper, J. Callaway and W. Armstrong, *J. Neurosci.* **24**, 4818 (2004).
5. R. K. S. Wong and D. A. Prince, *J. Neurophysiol.* **45**, 86 (1981).

6. V. Crunelli, J. S. Kelly, N. Leresche and M. Pirchio, *J. Physiol. (Lond.)* **384**, 587 (1987).
7. J. Rinzel, Bursting oscillations in an excitable membrane model, in *Ordinary and Partial Differential Equations*, eds. B. D. Sleeman and R. J. Jarvis, Lecture Notes in Mathematics, Vol. 1151 (Springer, Berlin, 1985).
8. J. Rinzel, A formal classification of bursting mechanisms in excitable systems, in *Mathematical Topics in Population Biology, Morphogenesis and Neurosciences*, eds. E. Teramoto and M. Yamaguti, Lecture Notes in Biomathematics, Vol. 71 (Springer–Verlag, Berlin, 1987).
9. R. Bertram, M. J. Butte, T. Kiemel and A. Sherman, *Bull. Math. Biol.* **57**, 413 (1995).
10. E. M. Izhikevich, *Int. J. Bifur. Chaos* **10**, 1171 (2000).
11. F. V. Goor, D. Zivadinovic, A. J. Martinez-Fuentes and S. S. Stojilkovic, *J. Biol. Chem.* **276**, 33840 (2001).
12. T. A. Kinard, G. de Vries, A. Sherman and L. S. Satin, *Biophys. J.* **76**, 1423 (1999).
13. R. Bertram and A. Sherman, Negative calcium feedback: The road from Chay-Keizer, in *Bursting: The Genesis of Rhythm in the Nervous System*, eds. S. Coombes and P. C. Bressloff (World Scientific, Singapore, 2005) pp. 19–48.
14. J. Tabak, N. Toporikova, M. E. Freeman and R. Bertram, *J. Comput. Neurosci.* **22**, 211 (2007).
15. A. P. LeBeau, A. B. McKinnon and J. Sneyd, *J. Theor. Biol.* **192**, 319 (1998).
16. J. V. Stern, H. M. Osinga, A. LeBeau and A. Sherman, *Bull. Math. Biol.* **70**, 68 (2008).
17. N. Toporikova, J. Tabak, M. E. Freeman and R. Bertram, *Neural Comput.* **20**, 436 (2008).

THE DYNAMICS OF ANTIBODY DEPENDENT ENHANCEMENT IN MULTI-STRAIN DISEASES WITH VACCINATION

LORA BILLINGS*

Department of Mathematical Sciences, Montclair State University,
Montclair, NJ 07043, USA
**E-mail: billingsl@mail.montclair.edu*
www.montclair.edu

IRA B. SCHWARTZ

US Naval Research Laboratory, Code 6792, Nonlinear System Dynamics Section,
Washington, DC 20375, USA
E-mail: ira.schwartz@nrl.navy.mil

LEAH B. SHAW

Department of Applied Science, College of William and Mary,
Williamsburg, VA 23187, USA
E-mail: lbshaw@wm.edu

In this paper, we present a model for multi-strain diseases with antibody-dependent enhancement (ADE) with single strain vaccination. The phenomenon of ADE can be described as the increase in the viral growth rate in a secondary infection after recovery from a primary infection of a different disease strain. Using center manifold analysis, a relation between the primary and secondary infection classes is derived on a lower dimensional space. We then examine the effect of single strain vaccination on the dynamics.

Keywords: epidemics; antibody-dependent enhancement; center manifold; vaccine.

1. Introduction

Some multi-strain diseases exhibit antibody-dependent enhancement (ADE), a phenomenon in which viral replication is increased rather than decreased by immune sera. Therefore, infectivity is increased during a second infection of the disease. The importance of understanding ADE is underscored by the pandemic status of the multi-strain disease dengue, which exhibits up to four serotypes. (For notation purposes, we consider serotypes and strains interchangeably in this work.) Dengue and other viruses have been shown to exhibit ADE in vitro.[1,2] Currently, there is no vaccine available to protect against all strains. Each year, tens of millions of dengue fever cases occur throughout the world. Additionally, hundreds of thousands of cases occur in a more severe form called dengue hemorrhagic fever (DHF).[3] Overall, DHF is reported to occur more frequently among individuals experiencing a secondary infection, since primary infections occur with mild or no symptoms.[4]

Because a dengue vaccine is not expected to become available for at least 5 to 10 years, the CDC urges the development of a surveillance system that provides early warning of an impending dengue epidemic.[3] Predicting the dynamics of a multi-strain disease with ADE could help in controlling the spread of dengue. Foundational work on ADE models can be found in Refs. 2,5–7. Similarly, there have been other approaches used in dengue models without ADE.[8,9] Following Ref. 10, we describe the lower-dimensional center manifold approximation of a model with a general number of co circulating strains and ADE factor, based on the ADE model of Refs. 6,7. We also investigate the effects of a single strain vaccine campaign, building on the work of Ref. 11. While we use parameters for dengue, the model is general enough to be considered for other multi-strain diseases.

2. Description of the General n-strain Model

We begin by describing the important characteristics of dengue epidemiology that we wish to model. Primary infection with any one strain confers immunity to that strain, but not to the others. Since tertiary infections are rare, we assume that individuals are immune to all four strains after two sequential strain infections.[4] It is also hypothesized that secondary infections carry a higher viral load, causing that person to be more infectious through ADE.[3] We design our model to use these properties, but for general n strains. Therefore, we follow the n-strain susceptible-infected-recovered (SIR) model formulation similar to the one derived in Ref. 7.

We use a compartmental model, in which the variables represent fractions of the total. Dengue is transmitted by mosquito, but the time scale for transmission is sufficiently short and the mosquito density is sufficiently dense to model the transmission by a mass action term, approximating it as person to person. The variable definitions are as follows:

s	Susceptible to all strains;
x_i	Primary infectious with strain i
r_i	Primary recovered from strain i
x_{ij}	Secondary infectious, currently infected with strain j, but previously had i $(i \neq j)$.

The model is a system of n^2+n+1 ordinary differential equations describing the rates of change of the population within each compartment:

$$\frac{ds}{dt} = \mu - \beta s \sum_{i=1}^{n} \left(x_i + \phi_i \sum_{j \neq i} x_{ji} \right) - \mu_d s \tag{1}$$

$$\frac{dx_i}{dt} = \beta s \left(x_i + \phi_i \sum_{j \neq i} x_{ji} \right) - \sigma x_i - \mu_d x_i \tag{2}$$

$$\frac{dr_i}{dt} = \sigma x_i - \beta r_i \sum_{j \neq i} \left(x_j + \phi_j \sum_{k \neq j} x_{kj} \right) - \mu_d r_i \tag{3}$$

$$\frac{dx_{ij}}{dt} = \beta r_i \left(x_j + \phi_j \sum_{k \neq j} x_{kj} \right) - \sigma x_{ij} - \mu_d x_{ij}. \tag{4}$$

The parameters μ, μ_d, β, and σ denote birth, death, contact, and recovery rates, respectively. The ADE effect is modeled by the parameters ϕ_i. We assume that the ADE factors are equal for all the strains in this analysis and use $\phi_i = \phi$, and $\phi \geq 1$. When $\phi = 1$, there is no ADE, and both primary and secondary infectives are equally infectious. When $\phi > 1$, the ADE appears as an enhancement factor in the nonlinear terms involving secondary infectives.

The dynamics of Eqs. 1–4 have been studied previously in Refs. 7 and 11. The known steady states are the disease free and the endemic equilibria. The disease free equilibrium (DFE) is the trivial case where no disease is present. The entire population is susceptible and all other compartments are empty. Its stability is determined by the basic reproduction number, or the maximum of the spectrum of the associated next generation matrix.[12] We can define the basic reproduction number for the system as

$$R_0 = \frac{\beta}{\mu + \sigma}. \tag{5}$$

When $R_0 > 1$, the DFE is unstable. Note that this value does not depend on the ADE factor. There are other equilibria that represent the die out of one or more strains, but due to the symmetry of the contact rate, they will never be stable. The dynamics are qualitatively similar as we vary n, the number of strains.

For $\phi \in [1, \phi_c)$, the system has a stable endemic steady state. At a critical ϕ value, ϕ_c, the system undergoes a Hopf bifurcation and begins to oscillate periodically. For slightly larger ϕ values, the periodic solutions become unstable and the system oscillates chaotically. Chaotic oscillations persist for larger values for ϕ, with the exception of narrow windows of stable periodic solutions.[7]

3. Lower Dimensional Center Manifold Approximation

Although the full n-strain model has n^2+n+1 dimensions, center manifold analysis shows the existence of an attractor in $2n+1$ dimensions, due to a relationship between primary and secondary infectives. Following Ref. 10, we consider Eqs. 1-4 with two strains ($n = 2$) and no mortality ($\mu_d = 0$). We scale the larger parameters β, σ by the small parameter μ, defining $\beta = \beta_0/\mu$ and $\sigma = \sigma_0/\mu$, where β_0, σ_0 are $\mathcal{O}(1)$. Therefore, we find the endemic steady state

$$s_0 = \frac{\sigma_0}{\beta_0(\phi+1)}, x_{i,0} = \frac{\mu^2}{n\sigma_0}, r_{i,0} = \frac{\sigma_0}{\beta_0(n-1)(\phi+1)}, x_{ij,0} = \frac{\mu^2}{n(n-1)\sigma_0}$$

for all $i, j = 1, 2$. We then shift the variables so that the endemic fixed point is at the origin: $\bar{s} = s - s_0, \bar{x}_i = x_i - x_{i,0}, \bar{r}_i = r_i - r_{i,0}, \bar{x}_{ij} = x_{ij} - x_{ij,0}$. By applying center manifold theory[13] to a combination of the transformed variables, we obtain the following approximation for the invariant manifold onto which the system collapses:

$$\begin{aligned}\sigma_0(\bar{x}_1 - \bar{x}_{21}) &= \beta_0 \left[\bar{s} - \bar{r}_2 + \frac{2-\phi}{1+\phi}\bar{x}_2 + \frac{3\phi}{1+\phi}\bar{x}_{12}\right](\bar{x}_1 + \phi\bar{x}_{21}),\\ \sigma_0(\bar{x}_2 - \bar{x}_{12}) &= \beta_0 \left[\bar{s} - \bar{r}_1 + \frac{2-\phi}{1+\phi}\bar{x}_1 + \frac{3\phi}{1+\phi}\bar{x}_{21}\right](\bar{x}_2 + \phi\bar{x}_{12}). \qquad (6)\end{aligned}$$

To bring out more of the structure, we make the following observations. We generally observed in numerical simulations that the infective compartments (and their deviations from the fixed point) are small compared to the susceptibles and recovereds. Thus the infective correction terms to $\bar{s} - \bar{r}_i$ may be dropped to obtain a simpler expression for the center manifold:

$$\begin{aligned}\sigma_0(\bar{x}_1 - \bar{x}_{21}) &= \beta_0 (\bar{s} - \bar{r}_2)(\bar{x}_1 + \phi\bar{x}_{21}),\\ \sigma_0(\bar{x}_2 - \bar{x}_{12}) &= \beta_0(\bar{s} - \bar{r}_1)(\bar{x}_2 + \phi\bar{x}_{12}). \qquad (7)\end{aligned}$$

The center manifold technique may again be applied for $n > 2$ strains. In the expressions for the center manifold, it is convenient to define the sum of secondary infectives currently infected with strain k: $\bar{z}_k = \sum_{i=1,i\neq k}^{n} \bar{x}_{ik}$. The following equations for the center manifold summarize our results for $n = 2, 3, 4$ and extrapolate them for general n:

$$\sigma_0 \left[\bar{x}_k - \bar{z}_k\right] = \beta_0[\bar{s} - \sum_{i\neq k} \bar{r}_i] \left(\bar{x}_k + \phi\bar{z}_k\right),$$

$$\sigma_0 \left[(n-1)\bar{x}_{jk} - \bar{z}_k\right] = \beta_0[(n-1)\bar{r}_j - \sum_{i\neq k} \bar{r}_i] \left(\bar{x}_k + \phi\bar{z}_k\right). \tag{8}$$

For each of the n strains, Eqs. 8 provide $n-1$ linearly independent equations, allowing $n(n-1)$ dimensions to be eliminated. Thus the dynamics of the system have dimension $2n+1$. If the susceptible and primary recovered compartments are known, a single quantity $\bar{z}_k$ for each strain is sufficient to describe the primary infective dynamics for that strain. Primary infectives can then easily be computed using Eqs. 8.

4. Vaccinations

We now analyze the effect of vaccinating against one strain on a population with four co-circulating strains ($n = 4$). We assume that young children are the targets of vaccination programs, which are a fraction of the new susceptibles entering the model at a rate of μ. We also assume that the vaccine has full efficacy and acts exactly like natural immunity. This is in contrast to the imperfect vaccination studied in Ref. 14, where a vaccinated individual could still contract two subsequent infections. Without loss of generality, we assume a fraction (v) are vaccinated against strain one, and modify the original model in the following way:

$$\frac{ds}{dt} = (1-v)\mu - \beta s \sum_{i=1}^{n} \left(x_i + \phi_i \sum_{j\neq i} x_{ji} \right) - \mu_d s,$$

$$\frac{dr_1}{dt} = v\mu + \sigma x_1 - \beta r_1 \sum_{j\neq 1} \left(x_j + \phi_j \sum_{k\neq j} x_{kj} \right) - \mu_d r_1. \tag{9}$$

If we assume $\mu_d = \mu$, there is mortality and we can analyze the stability of the modified DFE, or the solution with zeros for all of the infected components. The eigenvalues of the of the linearized system about the modified DFE are $\lambda_1 = \beta(1 - v - 1/R_0)$, $\lambda_2 = \beta(1 - v - 1/R_0 + \phi v)$, $\lambda_3 = -\mu$, and $\lambda_4 = -\mu - \sigma$, with multiplicities 1, 3, 5, and 12, respectively. For the

DFE to be stable, all of the eigenvalues would need negative real parts. We assume that in the absence of vaccination $R_0 > 1$, which is why there is a need to vaccinate. Therefore, $1/R_0 < 1$. Even if we had full vaccination ($v = 1$), we find $\lambda_2 = \beta(\phi - 1/R_0) > 0$ because the ADE factor $\phi > 1$ by definition.

To simplify analysis when the disease persists, we approximate the system using $\mu_d = 0$, which neglects mortality. For parameters that generate endemic oscillations ($\phi > \phi_c$), increasing the vaccination rate reduces the oscillations until the endemic steady state becomes stable through a reverse Hopf bifurcation. We show a sample time series of the effects of a vaccination campaign for $\phi = 3$ in Fig. 1. The parameters for the simulation are $\beta = 400$, $\sigma = 100$, and $\mu = 0.02$, all with units 1/year. After $t = 20$, we vaccinate at a rate of $v = 0.5$. You can see the amplitudes decreasing as the total number of infected approach an endemic state which is lower than the mean prior to vaccinations.

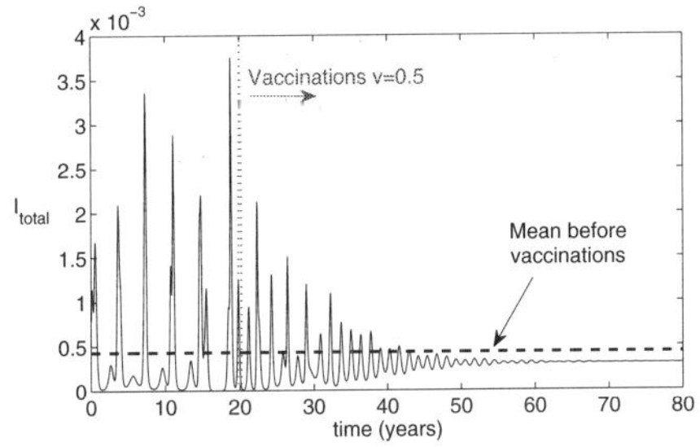

Fig. 1. Time series of total number of infectives before and after a single strain vaccine is implemented. See text for parameters.

Stabilizing the endemic state can be generalized over a range of ADE. For values of v just beyond the Hopf bifurcation, all four strains coexist, and asymptote to the steady state described by the following solution:

$$s = \frac{\sigma\,(1-v)}{\beta\,(1-v+\phi)}, \quad x_1 = \frac{(3\,v^2 - 4\,v + 1)\,\mu}{\sigma\,(4-3\,v)},$$

$$x_2 = x_3 = x_4 = \frac{\mu\,(1-v)}{\sigma\,(4-3\,v)},$$

$$r_1 = r_2 = r_3 = r_4 = \frac{\sigma}{3\beta\,(1-v+\phi)},$$

$$x_{i,1} = \frac{(1-3\,v)\,\mu}{3\sigma\,(4-3\,v)}, \quad x_{i,j(j\neq 1)} = \frac{\mu}{3\sigma\,(4-3\,v)}.$$

The dominant eigenvalue for the system linearized about this steady state is $\lambda_c = \frac{\sigma\phi(1-3v)}{2(1-v+\phi)}$. The solution is stable for $\lambda_c < 0$, or $v < 1/3$.

For $v > 1/3$, strain one dies out through a transcritical bifurcation and the following boundary equilibrium becomes stable:

$$s = \frac{\sigma\,(1-v)}{\beta\,(1-v+\phi)}, \quad r_1 = \frac{\sigma v}{\beta\,(1-v+\phi)},$$
$$x_1 = x_{2,1} = x_{3,1} = x_{4,1} = 0,$$
$$x_2 = x_3 = x_4 = \frac{\mu\,(1-v)}{3\sigma}, \quad r_2 = r_3 = r_4 = \frac{\sigma(1-v)}{2\beta\,(1-v+\phi)},$$
$$x_{1,2} = x_{1,3} = x_{1,4} = \frac{\mu v}{3\sigma}, \quad x_{i,j(i\neq 1, j\neq 1)} = \frac{\mu(1-v)}{6\sigma}.$$

Both endemic equilibria, for $x_1 = 0$ and $x_1 > 0$, have the total infected sum $\mu(2-v)/\sigma$, so the total number of people infected varies smoothly with the vaccination parameter. The overall effect of single strain vaccination can be measured by examining the fraction of the total infective population under vaccine to that without vaccine. The ratio is given by $(2-v)/2$. We are also interested in approximating the total number of secondary infected for these equilibria, which stays at a constant level μ/σ and does not vary with the vaccination rate. By not creating more secondary infected as we vaccinate against a strain, we do not raise the incidence of DHF in the population.

We show the bifurcation diagram for the transitions through these states as a function of ϕ and v in Fig. 2. The remaining parameters are the same as in Fig. 1.

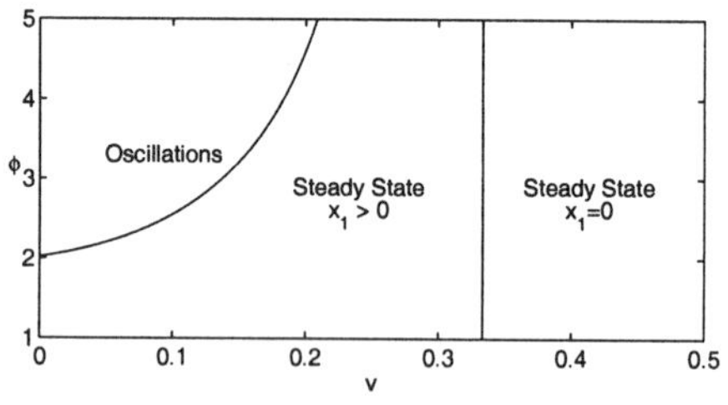

Fig. 2. The bifurcation curves for the vaccination model. See text for parameters.

5. Conclusions

In this paper, we have briefly examined the dynamics of a multi-strain disease model with antibody-dependent enhancement. We showed that by center manifold analysis we can derive a lower dimensional space on which the dynamics can be described. This reflects the organization between the primary and secondary infected classes. That is, we find that each strain exhibits time dependent synchrony between primary and secondary infections. Finally, we assessed the effect of a single-strain vaccination campaign. While the vaccination against one strain will not cause the other strains to die out, it will decrease the total number of infected people as the vaccination rate increases.

Acknowledgments

LB was supported by the National Science Foundation under Grant DMS-0414087. IBS was supported by the Office of Naval Research and the Armed Forces Medical Intelligence Center. LBS was supported by the Jeffress Memorial Trust.

References

1. D. S. Burke, *Perspect Biol Med* **35**, 511 (1992).
2. N. M. Ferguson, C. A. Donnelly and R. M. Anderson, *Phil. Trans. R. Soc. London, Ser. B* **354**, 757 (1999).
3. (2006), `http://www.cdc.gov/ncidod/dvbid/dengue/`.
4. A. Nisalak, T. P. Endy, S. Nimmannitya, S. Kalayanarooj, U. Thisayakorn, R. M. Scott, D. S. Burke, C. H. Hoke, B. L. Innis and D. W. Vaughn, *Am. J. Trop. Med. Hyg.* **68**, 191 (2003).
5. N. Ferguson, R. Anderson and S. Gupta, *Proc. Natl. Acad. Sci. USA* **96**, 790 (1999).
6. D. A. T. Cummings, I. B. Schwartz, L. Billings, L. B. Shaw and D. S. Burke, *Proc. Natl. Acad. Sci. USA* **102**, 15259 (2005).
7. I. B. Schwartz, L. B. Shaw, D. A. T. Cummings, L. Billings, M. McCrary and D. Burke, *Phys. Rev. E* **72**, art. no. (2005).
8. L. Esteva and C. Vargas, *Math. Biosciences* **167**, 51 (2000).
9. L. Esteva and C. Vargas, *J. Math. Biol.* **46**, 31 (2003).
10. L. B. Shaw, L. Billings and I. B. Schwartz, *J. Math. Biol.* **55**, 1 (2007).
11. L. Billings, I. B. Schwartz, L. B. Shaw, M. McCrary, D. Burke and D. A. T. Cummings, *J. Theor. Biol.* **246**, 18 (2007).
12. P. van den Driessche and J. Watmough, *Math. Biosciences* **180**, 29 (2002).
13. J. Carr, *Applications of centre manifold theory* (Springer-Verlag, New York, 1981).
14. L. Billings, A. Fiorillo and I. B. Schwartz, *Math. Biosciences* **211**, 265 (2008).

DYNAMICS OF INEXTENSIBLE VESICLES SUSPENDED IN A CONFINED TWO-DIMENSIONAL STOKES FLOW

A. RAHIMIAN, S. K. VEERAPANENI,

Department of Mechanical Engineering and Applied Mechanics
University of Pennsylvania,
Philadelphia, PA 19104, USA
email: rahimian@seas.upenn.edu *&* shravan@seas.upenn.edu

D. ZORIN

Courant Institute of Mathematical Sciences
New York University
New York, NY 10003, USA
email: dzorin@cs.nyu.edu

G. BIROS

Departments of Mechanical Engineering and Applied Mechanics,
Bioengineering, and Computer and Information Science,
University of Pennsylvania,
Philadelphia, PA 19104, USA
email: biros@seas.upenn.edu

The formulation and numerical solution of the dynamics of suspended vesicles in a bounded Stokes flow is outlined in this paper. We propose a semi-implicit time marching scheme, which is based on a boundary integral formulation of the corresponding hydrodynamic flow equations and vesicle deformation dynamics. We restrict our attention to the case in which the fluids inside and outside the vesicle have the same density and viscosity. We use a boundary integral formulation for the fluid that results in a set of nonlinear integro-differential equations for the vesicle dynamics. The motion of the vesicles is governed by the interplay between hydrodynamic and elastic forces. Fluid-structure interaction problems of this type are challenging to simulate. On one hand, explicit time-stepping schemes suffer from a severe stability constraint due to the stiffness related to high-order spatial derivatives. On the other hand, implicit time-stepping schemes can be expensive because they require the solution of a set of nonlinear equations at each time step. To circumvent these difficulties, we employ a semi-implicit scheme that does not have severe stability constraints and whose computational cost per time step is comparable to that of an explicit scheme. We discretize the equations by using a spectral method in space, and a first-order time stepping scheme in time. We report numerical experiments that demonstrate the convergence properties of our scheme.

Keywords: Interfacial dynamics; Stokes flow; Integral equations.

1. Introduction

Vesicles are closed lipid membranes suspended in a viscous (typically aqueous) solution. Vesicles are present in many biological phenomena[1,2] and are used experimentally to understand properties of biological membranes.[3] In addition, vesicle mechanics have been served as models for red blood cells.[4,5] Despite the importance of vesicle flows, however, the design of efficient computational methods for such problems has received limited attention.

In this article, we focus our attention on numerical schemes for continuum models of vesicle dynamics. This problem is challenging because the motion and shape of the vesicles are not given a priori, but determined dynamically from a balance of interfacial forces with fluid stresses. The shape dynamics of fluid vesicles is governed by the coupling of the flow within the two dimensional membrane of the vesicle to the hydrodynamics of the surrounding bulk fluid. We present a numerical scheme for the simulation of the motion of arbitrarily shaped vesicles in a confined geometry. In particular, we extend our past work in Ref. 6 by including rigid boundaries with prescribed boundary conditions on the velocity. We outline an integral equation formulation for the evolution of the vesicle. In contrast to stencil-based methods (e.g., finite element methods), integral equation formulations avoid discretization of the overall domain and instead discretize only the vesicle boundary and the boundary of the enclosing domain. This is the main reason that integral equations have been used extensively for

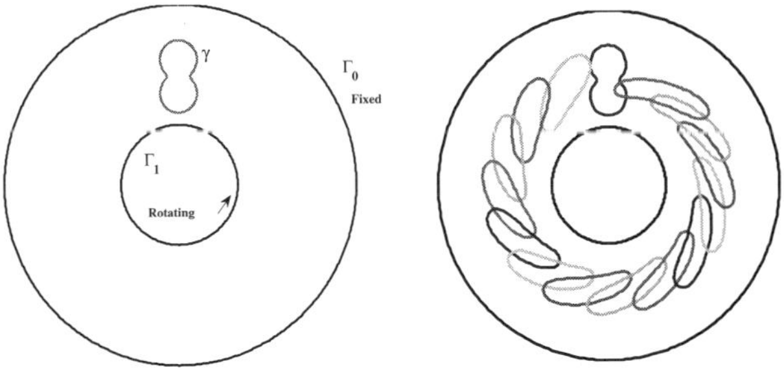

(a) Test problem.

(b) Evolution of the vesicle in time, T = 140s, (color coded in time) (color can be viewed from the e-version).

Fig. 1. Simulation of a single vesicle in concentric setup of Couette flow.

vesicle, and more generally, particulate and interfacial flow simulations.[7] Also, we present results from numerical experiments that demonstrate the stability of the proposed time marching scheme.

One significant challenge in simulating vesicle dynamics is the numerical stiffness of the integro-differential equations that describe these dynamics.[6] The overwhelming majority of work on particulate flows uses explicit schemes that pose severe restrictions on the time step. In problems with vesicles, the elastic and incompressibility properties of their membranes must be taken into account and the numerical schemes must be modified in order to solve the resulting set of boundary integral equations. Details of the boundary integral formulation for elastic interfaces and incompressible vesicles can be found in Refs. 7 and 8.

Limitations. The most significant limitation of our scheme is that it is not adaptive in time and space (the time step and spatial discretization must be selected manually by the user). In addition, we restrict our attention only to dilute suspensions of vesicles in fluids with bounded domains. Here, we do not consider forces due to gravity, electrostatics or adhesion, or inertial effects due to the mass of the fluid or the vesicle membrane. The method described in Ref. 9 was used to evaluate the double layer potential close to the boundary of the domain. Nevertheless, maintaining high accuracy for vesicles closely approaching each other requires incorporation of specialized quadrature rules for nearly-singular integrals, as well as appropriate models for short-range interaction forces and efficient collision-detection schemes.

Outline of the paper. In Section 2, we give a precise statement of the problem and its integral equation formulation. In Section 3, we outline the numerical scheme we use to solve the derived equations. We propose a computational scheme for the evolution of vesicles in confined domain. Our scheme is based on Lagrangian tracking of marker particles on the vesicle and a semi-implicit time discretization. High-order accuracy in space is ensured by using a Fourier basis discretization for all functions and computing derivatives in Fourier domain, as well as high-order, Gauss-trapezoidal quadrature rules[10] for discretization of the double-layer and single-layer potentials. In time, we use a semi-implicit marching scheme, first derived for advection-diffusion equations.[11] This discretization yields a linear system of equations for each time step, which is solved using a Krylov iterative method (GMRES).[12] Throughout this paper, scalars will be represented by lowercase italic and vectors by lowercase boldface letter. We use $\mathcal{C}[\varphi](\mathbf{x}) := \int_\gamma C(\mathbf{x},\mathbf{y})\varphi(\mathbf{y})\,ds(\mathbf{y})$ to denote the convolution of the kernel C with density φ; $s(\mathbf{y})$ is defined as a general parametrization of the curve γ and not specifically the arc-length parametrization. By $\otimes$, we denote the tensor product of two vectors. For vectors, numerical subscripts imply the components of the vector; numerical superscripts are employed to enumerate vectors within a set of vectors.

Acknowledgments. We would like to thank the organizers of the Conference on Frontiers in Applied and Computational Mathematics. This work was supported by the U.S. Department of Energy under grant DE-FG02-04ER25646, and the U.S. National Science Foundation grants DMS-0612578, OCI 0749285, and OCI 0749334.

2. Formulation

We consider the formulation for a *single* vesicle suspended in a two dimensional multiply connected domain Ω of fluid. (In Section 3.1, we generalize the formulation to the case of multiple vesicles.) Ω is a compact subset of $\mathbb{R}^2$ and its boundary consists of $n+1$ infinitely differentiable curves $\Gamma_0, \ldots, \Gamma_n$. The outer boundary of the region is Γ_0. Let $\Gamma = \bigcup_{k=0}^{n} \Gamma_k$ and let γ denote the boundary of the vesicle. Then, the motion of the fluid inside Ω is given by

$$\nabla P - \mu \Delta \mathbf{u} = \mathbf{f}; \quad \operatorname{div} \mathbf{u} = 0 \quad \text{in} \quad \Omega \backslash (\gamma \cup \Gamma). \tag{1}$$

$p(\mathbf{x})$ is the pressure field, $\mathbf{u}(\mathbf{x})$ is the velocity field in Ω, μ is the viscosity of the fluid, and $\mathbf{f}$ is the force imposed by the deformable vesicle membrane on the fluid (that is, a local force having a nonzero value only at the membrane interface). The no-slip boundary condition at the boundaries and vesicle interface requires that

$$\mathbf{u}(\mathbf{x}) = \dot{\mathbf{x}} \quad \text{for} \quad \mathbf{x} \in \gamma, \qquad \mathbf{u}(\mathbf{x}) = \mathbf{U}(\mathbf{x}) \quad \text{for} \quad \mathbf{x} \in \Gamma, \tag{2}$$

where $\dot{\mathbf{x}}$ is the velocity of points on the vesicle membrane and $\mathbf{U}$ is the prescribed velocity on the boundary of the domain.

Following Ref. 6, we use $\mathbf{f}_b$ and $\mathbf{f}_\sigma$ to denote the forces acting on γ due to bending and tension respectively. The tension σ acts as a Lagrange multiplier enforcing local inextensibility. One can show that $\mathbf{f}_\sigma = (\sigma \mathbf{x}_s)_s$ and $\mathbf{f}_b = -\mathbf{x}_{ssss}$, where the subscript s implies differentiation with respect to arc-length. The total force $\mathbf{f} = \mathbf{f}_b + \mathbf{f}_\sigma$ is balanced by the fluid traction jump across the vesicle membrane. Using potential theory,[13–15] we can reformulate Eqs. (1) and (2) as boundary integral equations. Due to linearity of the Stokes equations, the flow can be written as a superposition of two terms, namely the contribution from the vesicle boundary, plus a contribution due to the undisturbed applied Stokes flow in the domain. Hence, the dynamics of the vesicle are governed by

$$\mathbf{u}(\mathbf{x}) = \mathcal{S}[\mathbf{f}](\mathbf{x}) + \mathcal{D}[\boldsymbol{\eta}](\mathbf{x}), \quad \mathbf{x} \in \Omega \backslash (\gamma \cup \Gamma) \tag{3}$$

$$\mathbf{x}_s \cdot \dot{\mathbf{x}}_s = 0 \quad \mathbf{x} \in \gamma; \tag{4}$$

here $\boldsymbol{\eta}(\mathbf{x})$ is the double layer potential density over Γ (boundary of the domain) and $\mathbf{x}_s$ gives us the tangent vector. The second equation enforces the local inextensibility of the interface by requiring the surface divergence of the velocity field on the vesicle-fluid interface to vanish. The kernels of single and double layer potentials $\mathcal{S}[\mathbf{f}](\mathbf{x})$ and $\mathcal{D}[\boldsymbol{\eta}](\mathbf{x})$ are defined as

$$S(\mathbf{x}, \mathbf{y}) := \frac{1}{4\pi\mu}\left(-\ln \rho \, \mathbf{I} + \frac{\mathbf{r} \otimes \mathbf{r}}{\rho^2}\right), \tag{5}$$

$$D(\mathbf{x}, \mathbf{y}) := \frac{1}{\pi} \frac{\mathbf{r} \cdot \mathbf{n}(\mathbf{y})}{\rho^2} \frac{\mathbf{r} \otimes \mathbf{r}}{\rho^2}. \tag{6}$$

Here, $\mathbf{r} = \mathbf{x} - \mathbf{y}$, $\rho = \|\mathbf{r}\|_2$, and $\mathbf{n}(\mathbf{y})$ is the outward normal to Γ at $\mathbf{y}$.

As it is shown in Refs. 13 and 15, the double-layer operator is rank deficient when Ω is a multiply-connected domain. To remove the null space of the double-layer operator, we add *Stokeslet* and *Rotlet* terms for each Γ_k $(1 \leq k \leq n)$ to Eq. (3). The Stokeslet is the Green's function of the Stokes equation that is the same as the single layer kernel S given in Eq. (5). The Rotlet is the antisymmetric component of the Stokes doublet defined by

$$R(\mathbf{x}, \mathbf{y})(\boldsymbol{\xi}) := \frac{1}{\mu} \frac{\boldsymbol{\xi} \times \mathbf{r}}{\rho^2}, \tag{7}$$

for any strength vector $\boldsymbol{\xi}$. Here $\times$ denotes the cross product of vectors (in two dimensions, $\mathbf{r}$, of course, lies in the x-y plane and $\boldsymbol{\xi}$ is a vector with nonzero entry only in the z direction). Both the Stokeslet and Rotlet are centered at an interior point $\mathbf{c}^k$ of the domain enclosed by the boundary component Γ_k. For convenience we choose $\boldsymbol{\lambda}$, the strength vector for each Stokeslet, and $\boldsymbol{\xi}$ to depend linearly on the unknown density $\boldsymbol{\eta}$ in the following manner

$$\boldsymbol{\lambda}_1^k := \frac{1}{2\pi} \int_{\Gamma_k} \boldsymbol{\varphi}^1(\mathbf{y}) \cdot \boldsymbol{\eta}(\mathbf{y}) \, ds(\mathbf{y}), \ \boldsymbol{\lambda}_2^k := \frac{1}{2\pi} \int_{\Gamma_k} \boldsymbol{\varphi}^2(\mathbf{y}) \cdot \boldsymbol{\eta}(\mathbf{y}) \, ds(\mathbf{y}), \tag{8}$$

and

$$\boldsymbol{\xi}_1^k = \boldsymbol{\xi}_2^k = 0, \ \boldsymbol{\xi}_3^k := \frac{1}{2\pi} \int_{\Gamma_k} \boldsymbol{\varphi}^3(\mathbf{y}) \cdot \boldsymbol{\eta}(\mathbf{y}) \, ds(\mathbf{y}) \tag{9}$$

where $\boldsymbol{\varphi}^1 = \delta_{1i}, \boldsymbol{\varphi}^2 = \delta_{2i}$ $(i = 1, 2)$ are the rigid body transformation in the plane and $\boldsymbol{\varphi}^3$ is the rigid body rotation, that is $\boldsymbol{\varphi}^3(\mathbf{y}) = \{y_2, -y_1\}$.

Since the flow is confined by the contour Γ_0, as it is noted in Refs. 15 and 16, conservation of mass ($\mathrm{div}\mathbf{u} = 0$) implies that the velocity field defined by Eq. (3) will satisfy Eq. (1) only when $\boldsymbol{\eta}$ satisfies $\int_{\Gamma_0} \boldsymbol{\eta} \cdot \mathbf{n}\, ds = 0$. To enforce this orthogonality condition we follow Ref. 16 and add an additional operator $\mathcal{N}[\boldsymbol{\eta}](\mathbf{x}) = \int_{\Gamma_0} N(\mathbf{x},\mathbf{y})\boldsymbol{\eta}(\mathbf{y})\, ds(\mathbf{y})$ with kernel $N(\mathbf{x},\mathbf{y}) = \mathbf{n}(\mathbf{x}) \otimes \mathbf{n}(\mathbf{y})$ to the right hand side of Eq. (3). Hence, the flow in $\Omega\backslash(\gamma \cup \Gamma)$ can be represented by

$$\begin{aligned}\mathbf{u}(\mathbf{x}) &= \mathcal{S}[\mathbf{f}](\mathbf{x}) + \mathcal{D}[\boldsymbol{\eta}](\mathbf{x}) + \mathcal{N}[\boldsymbol{\eta}](\mathbf{x}) + \sum_{k=1}^{n} R(\mathbf{x},\mathbf{c}^k)\boldsymbol{\xi}^k + \sum_{k=1}^{n} S(\mathbf{x},\mathbf{c}^k)\boldsymbol{\lambda}^k,\\ &= \mathcal{S}[\mathbf{f}](\mathbf{x}) + \mathcal{Q}(\boldsymbol{\eta},\Xi,\Lambda)(\mathbf{x}),\end{aligned} \tag{10}$$

where $\Xi = \{\boldsymbol{\xi}^1,\ldots,\boldsymbol{\xi}^n\}$, $\Lambda = \{\boldsymbol{\lambda}^1,\ldots,\boldsymbol{\lambda}^n\}$, and

$$\mathcal{Q} = \mathcal{D}[\boldsymbol{\eta}](\mathbf{x}) + \mathcal{N}[\boldsymbol{\eta}](\mathbf{x}) + \sum_{k=1}^{n} R(\mathbf{x},\mathbf{c}^k)(\boldsymbol{\xi}^k) + \sum_{k=1}^{n} S(\mathbf{x},\mathbf{c}^k)(\boldsymbol{\lambda}^k). \tag{11}$$

In this way, we obtain a system of Fredholm integral equations of the second kind. Taking the limit of Eq. (10) for points $\boldsymbol{x}$ on the boundary of the domain, we obtain an equation for $\boldsymbol{\eta}$ supplemented with Eq. (8) and Eq. (9) for calculation of $\boldsymbol{\lambda}^k$ and $\boldsymbol{\xi}^k$. At the vesicle boundary the velocity is equal to that of the adjacent fluids.

Thus, from Eq. (10) combined with the inextensibility condition on the membrane, we obtain a set of equations that fully describe the evolution of the vesicle:

$$\dot{\mathbf{x}} = \mathcal{S}[\mathbf{f}](\mathbf{x}) + \mathcal{Q}(\boldsymbol{\eta},\Xi,\Lambda)(\mathbf{x}) \qquad \mathbf{x} \in \gamma, \tag{12a}$$

$$\mathbf{x}_s \cdot \dot{\mathbf{x}}_s = 0 \qquad \mathbf{x} \in \gamma, \tag{12b}$$

$$U(\mathbf{x}) = \mathcal{S}[\mathbf{f}](\mathbf{x}) + \mathcal{Q}(\boldsymbol{\eta},\Xi,\Lambda)(\mathbf{x}) - \frac{1}{2}\boldsymbol{\eta}(\mathbf{x}) \quad \mathbf{x} \in \Gamma, \tag{12c}$$

$$\boldsymbol{\lambda}^k = \frac{1}{2\pi}\int_{\Gamma_k} \boldsymbol{\eta}(\mathbf{y})\, ds(\mathbf{y}), \tag{12d}$$

$$\xi_3^k = \frac{1}{2\pi}\int_{\Gamma_k} \boldsymbol{\varphi}^3(\mathbf{y}) \cdot \boldsymbol{\eta}(\mathbf{y})\, ds(\mathbf{y}). \tag{12e}$$

Given the density $\boldsymbol{\eta}$ on the boundary of Ω, the strengths of Stokeslets and Rotlets, Ξ and Λ, and the traction jump across the vesicle, we can calculate the velocity at any point $\mathbf{x} \in \Omega\backslash(\Gamma \cup \gamma)$ by Eq. (10). Thus, Eqs. (12) and (10) represent the reformulation of Eq. (1), Eq. (2), and the force balance across the interface as a system of integro-differential equations.

3. Numerical scheme

In this section, we present the proposed numerical scheme to solve the system of equations (12). First, we discuss the discretization in time and second, the discretization in space. There are several ways to represent and track smooth interfaces.

Membrane evolution. Let $\alpha \in [0, 2\pi)$ be the parametrization of the boundary γ and let $\mathbf{x}^n(\alpha)$ be the position of the interface at the nth time step. A first-order semi-implicit scheme to compute the position at the $(n+1)$th step is given by

$$\begin{aligned} &\frac{1}{\Delta t}(\mathbf{x}^{n+1} - \mathbf{x}^n) = \mathcal{S}[\mathbf{f}^{n+1}](\mathbf{x}^n) + \mathcal{Q}(\boldsymbol{\eta}^n, \Xi^n, \Lambda^n)(\mathbf{x}^n) \\ &\mathbf{x}^n_\alpha \cdot \left(\mathcal{S}[\mathbf{f}^{n+1}](\mathbf{x}^n) + \mathcal{Q}(\boldsymbol{\eta}^n, \Xi^n, \Lambda^n)(\mathbf{x}^n)\right)_\alpha = 0, \end{aligned} \tag{13}$$

for $\mathbf{x} \in \gamma(\alpha, t)$, where

$$\mathcal{S}[\mathbf{f}^{n+1}](\mathbf{x}^n) = \int_0^{2\pi} S(\mathbf{x}^n, \mathbf{y}^n)\mathbf{f}^{n+1} d\alpha, \tag{14}$$

$$\mathbf{f}^{n+1} = \left(\frac{1}{|\mathbf{y}^n_\alpha|}\left(\frac{1}{|\mathbf{y}^n_\alpha|}\left(\frac{\mathbf{y}^{n+1}_\alpha}{|\mathbf{y}^n_\alpha|}\right)_\alpha\right)_\alpha\right)_\alpha + \left(\sigma^{n+1}\frac{\mathbf{y}^n_\alpha}{|\mathbf{y}^n_\alpha|}\right)_\alpha. \tag{15}$$

Note that the Jacobian $|\mathbf{x}_\alpha|$, due to local inextensibility, is time independent. This motivates its explicit treatment. The tension and the term in the bending force with highest derivative were treated implicitly. The rest of the terms are all treated explicitly. The high-order quadrature rules given in Ref. 10, designed to handle the logarithmic singularity, are used to compute the integrals. The unknowns are the position $\mathbf{x}^{n+1}$ and tension σ^{n+1}. The system of coupled equations (13) is linear in the unknowns and solved using GMRES.[12]

To evaluate the double-layer density over the boundary of the domain Γ, we need to solve Eqs. (12c)-(12e). For points on Γ, the double-layer kernel $D(\mathbf{x}, \mathbf{y})$ has no singularities; for $\mathbf{x}, \mathbf{y} \in \Gamma$ $\lim_{\mathbf{y}\to\mathbf{x}} D(\mathbf{x}, \mathbf{y}) = -\frac{\kappa}{2\pi}\, \mathbf{t} \otimes \mathbf{t}$. We discretize Eqs. (12c)-(12e) by the Nyström method combined with composite trapezoidal rule to achieve superalgebraic convergence for smooth data. For any $\mathbf{x}$ on Γ we have

$$\begin{aligned} &\mathbf{U}(\mathbf{x}) - \mathcal{S}[\mathbf{f}^{n+1}](\mathbf{x}) = -\frac{1}{2}\boldsymbol{\eta}^{n+1} + \mathcal{Q}(\boldsymbol{\eta}^{n+1}, \Xi^{n+1}, \Lambda^{n+1})(\mathbf{x}), \\ &(\boldsymbol{\lambda}^k)^{n+1} = \frac{1}{2\pi}\int_{\Gamma_k} \boldsymbol{\eta}^{n+1}(\mathbf{y})\, ds(\mathbf{y}),\ (\xi^k_3)^{n+1} = \frac{1}{2\pi}\int_{\Gamma_k} \boldsymbol{\varphi}^3(\mathbf{y}) \cdot \boldsymbol{\eta}^{n+1}(\mathbf{y})\, ds(\mathbf{y}). \end{aligned} \tag{16}$$

This gives us a dense system of equations. Due to explicit treatment of the flow, Eq. (16) is decoupled with Eq. (13) and solved separately at each time

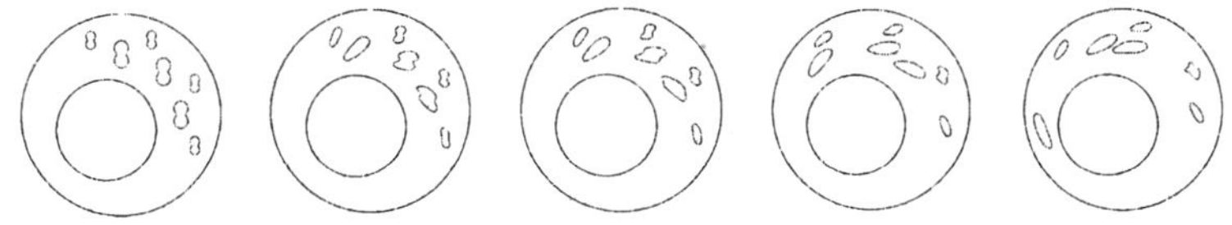

(a) $t = 0$s. (b) $t = 10$s. (c) $t = 20$s. (d) $t = 40$s. (e) $t = 80$s.

Fig. 2. Simulation of a multiple vesicle in eccentric setup of Couette flow.

step. In summary, at time step $n+1$, we first solve the coupled system of equations (13) for $\mathbf{x}^{n+1}$ and $\boldsymbol{\sigma}^{n+1}$. Then, we solve Eqs. (16) for $\boldsymbol{\eta}^{n+1}, \Lambda^{n+1}$, and Ξ^{n+1}.

3.1. *Multiple vesicles*

If K vesicles are suspended in the domain, Eq. (12b) can be expanded into the following equation for the evolution of the jth vesicle

$$\dot{\mathbf{x}}_j = \mathcal{Q}(\boldsymbol{\eta}, \Xi, \Lambda)(\mathbf{x}_j) + \sum_{k=1}^{K} \mathcal{S}[\mathbf{f}_k](\mathbf{x}_j), \tag{17}$$

here $\mathcal{S}[\mathbf{f}_k](\mathbf{x}_j)$ denotes the single layer potential evaluated over the kth vesicle. Following the same procedure as above and treating the effect of vesicles on each other explicitly, we can write the discrete form of Eq. (17) as

$$\begin{aligned}\frac{1}{\Delta t}(\mathbf{x}_j^{n+1} - \mathbf{x}_j^n) &= \mathcal{Q}(\boldsymbol{\eta}^n, \Xi^n, \Lambda^n)(\mathbf{x}_j^n) \\ &\quad + \mathcal{S}[\mathbf{f}_j^{n+1}](\mathbf{x}_j^n) + \sum_{\substack{k=1 \\ k\neq j}}^{n} \mathcal{S}[\mathbf{f}_k^n](\mathbf{x}_j^n),\end{aligned} \tag{18}$$

supplemented with the inextensibility constraint

$$(\mathbf{x}_j^{n+1})_\alpha \cdot (\dot{\mathbf{x}}_j^{n+1})_\alpha = 0. \tag{19}$$

The number of vesicles has no effect on the evaluation of the density over Γ and we follow exactly the same procedure as in the case of a single vesicle.

4. Results

The stability and convergence properties of the proposed numerical time-stepping scheme for the evolution of vesicle in an unbounded domain were discussed in Ref. 6. Here, we focus on the accuracy of the model by comparing the position of the vesicle with the reference case. Figure 1(a) shows

Table 1. The error in position X_h, length L_h, and area A_h in the evolution of a single vesicle in a concentric Couette flow. The error in the position is calculated by comparing to a geometry obtained with finer resolution ($M = 256$). Due to inextensibility and incompressibility, the area and length should be preserved. We set $\Delta t = \frac{T}{aM}$, where T is the time horizon and a is an arbitrary constant(we choose $a = 5$).

T	Number of points on Γ	Number of points on γ	$\frac{\lvert X - X_h \rvert_\infty}{\lvert X \rvert_\infty}$	$\frac{\lvert L - L_h \rvert}{L}$	$\frac{\lvert A - A_h \rvert}{A}$
	32	32	2.12e+00	4.80e−02	2.70e−02
80s	64	64	9.57e−01	2.49e−02	1.30e−02
	128	128	3.28e−01	1.27e−02	6.33e−03

Table 2. The error in position X_h, length L_h, and area A_h of the evolution of multiple vesicles in an eccentric Couette flow. The errors are calculated by comparing to the "exact" solution ($M = 256$). We set $\Delta t = \frac{T}{aM}$, where T is the time horizon and a is an arbitrary constant (we choose $a = 2$).

T	Number of points on Γ	Number of points on each vesicle γ_j	$\frac{\lvert X - X_h \rvert_\infty}{\lvert X \rvert_\infty}$	$\frac{\lvert L - L_h \rvert}{L}$	$\frac{\lvert A - A_h \rvert}{A}$
	32	32	6.75e-02	2.61e-02	3.80e-01
10s	64	64	1.05e-02	1.69e-02	3.54e-03
	128	128	3.85e-03	1.30e-02	2.19e-03

a vesicle in a bounded domain of two concentric circles. In our test case, we fix the outer circle (fluid velocity zero) and we impose a clock-wise rotation on the inner circle. Our setup models a 2D Couette flow. The initial shape of the vesicle is biconcave. Figure 1(b) shows the time evolution of the vesicle. The exact solution of Eqs. (18) and (16) is not known analytically. To test our solver, we obtain a highly accurate solution using a large number of discretization points in space and time then we use it to calculate the error for much coarser discretizations. We report the results of this experiments in Table 1.

As an additional test case, we consider the dynamics of multiple vesicle in the case of an eccentric Couette flow. The test domain is shown in Fig. 2(a). The position of the vesicles in different time frames is given in Figs. 2(b)-2(e). We report the error in the final position and the maximum error in the area and length of the vesicles among all vesicles in Table 2. The

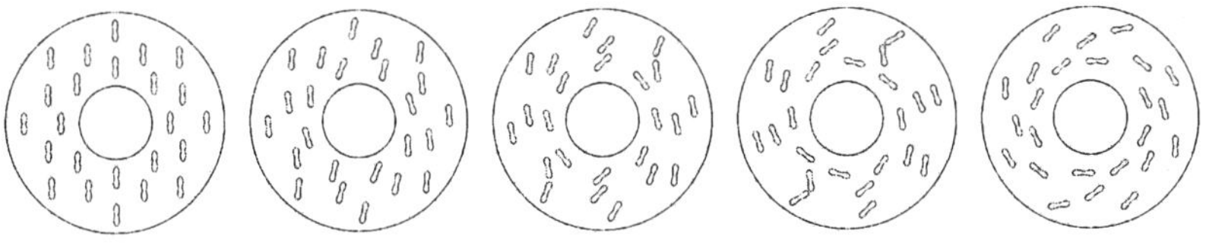

(a) $t = 0$s. (b) $t = 10$s. (c) $t = 26$s. (d) $t = 45$s. (e) $t = 90$s.

Fig. 3. Simulation of a twenty four vesicles in Couette flow, $\kappa = 1$.

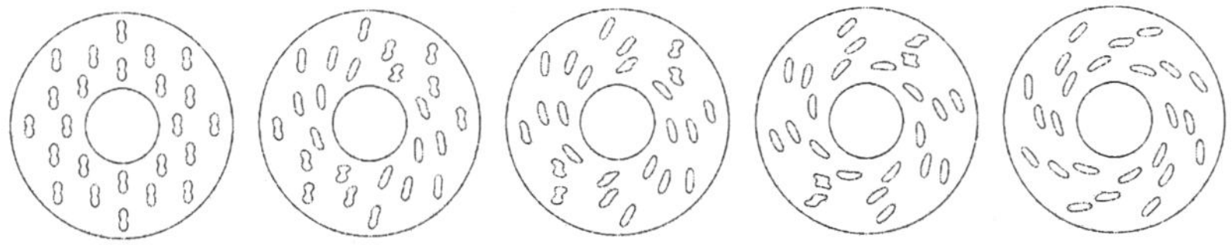

(a) $t = 0$s. (b) $t = 10$s. (c) $t = 20$s. (d) $t = 40$s. (e) $t = 80$s.

Fig. 4. Simulation of a twenty four vesicles in Couette flow, $\kappa = 0.01$.

area and the length of the vesicle is conserved due to its inextensibility and incompressibility. This number gives us a good measure of the accuracy of the solver. The semi-implicit treatment of the problem and the relatively smaller number of required points for numerical solution of integral equations compared to other grid based method give us the freedom to solve larger problems and carry on long-time simulations. As an example Figs. 3 and 4 show snapshots of the time evolution of twenty four vesicles in the Couette flow. We discretized using 32 points on each vesicle and 128 points on the boundary. We have selected $dt = 0.5$s, the time step for this simulation. We observe fist-order accuracy, which is expected since we are using a first-order accurate scheme in time.

5. Conclusions

In this article, we presented a semi-implicit scheme to model the evolution of vesicles in a confined domain and convergence results have been presented. The present study has been concerned with a 2D problem. It has been found that the basic features in 3D are similar to those captured in 2D simulations.[17–19] These various studies seem to point out that the 2D model captures essential features. Currently, we are working in extending the numerical scheme to the 3D case.

References

1. M. Kraus, W. Wintz, U. Seifert and R. Lipowsky, *Phys. Rev. Lett.* **77**, 3685(Oct 1996).
2. U. Seifert, *Advances in Physics* **46**, 13 (1997).
3. E. Sackmann, *Science* **271**, 43 (1996).
4. H. Noguchi and D. G. Gompper, *Proceedings Of The National Academy Of Sciences Of The United States Of America* **102**, 14159 (2005).
5. C. Pozrikidis, *Journal of Fluid Mechanics* **216**, 231 (1990).
6. S. K. Veerapaneni, G. Biros, D. Gueyffier and D. Zorin, *A boundary integral method for simulating the dynamics of inextensible vesicles suspended in a viscous fluid in 2d*, tech. rep., University of Pennsylvania (2008), Submitted for publication.
7. C. Pozrikidis, *Journal of Computational Physics* **169**, p. 250301 (2001).
8. C. Pozrikidis, *Journal of Fluid Mechanics* **297**, p. 123152 (1995).
9. L., G. Biros and D. Zorin, *J. Comput. Phys.* **219**, 247 (2006).
10. B. K. Alpert, *SIAM Journal on Scientific Computing* **20**, 1551 (1999).
11. U. M. Ascher, S. J. Ruuth and B. T. R. wetton, *SIAM Journal on Numerical Analysis* **32**, 797 (1995).
12. Y. Saad, *Iterative Methods for Sparse Linear Systems*, second edn. (Society for Industrial and Applied Mathematics, Philadelphia, PA, USA, 2003).
13. H. Power, *IMA J Appl Math* **51**, 123 (1993).
14. H. Power and G. Miranda, *SIAM J. Appl. Math.* **47**, 689 (1987).
15. C. Pozrikidis, *Boundary Integral and Singularity Methods for Linearized Viscous Flow* (Cambridge University Press, New York, NY, USA, 1992).
16. S. J. Karrila and S. Kim, *Chemical engineering communications* **82**, 123 (1989).
17. S. Sukumaran and U. Seifert, *Phys. Rev. E* **64**, p. 011916(Jun 2001).
18. V. Kantsler and V. Steinberg, *Physical Review Letters* **96**, p. 036001 (2006).
19. J. Beaucourt, F. Rioual, T. Séon, T. Biben and C. Misbah, *Phys. Rev. E* **69**, p. 011906(Jan 2004).

A ROBUST RECURSIVE PARTITIONING ALGORITHM FOR MINING MULTIPLE POPULATIONS

J. ALVIR[1], J. CABRERA[2], F. CARIDI[1], H. NGUYEN[1] and C. ROBERTS[1]
[1]*Pfizer Inc., New York, NY,*
[2]*Rutgers U., Piscataway, NJ, E-mail: cabrera@rutgers.edu*

One of the most important questions in the drug industry is to be able to characterize patients who benefit the most from a drug compared to older drugs or to competitive drugs. In the simplest setting we have two treatments A and B and we would like to characterize the patients who benefit from drug A more than from drug B. We measure this benefit with a statistic that computes the difference of mean treatment effect. Standard data mining techniques like recursive partitioning (CART, C4.5) optimize a criterion for partitions of a single population, but in this case we deal with criteria that evaluates partitions of multiple populations. In this paper we introduce a new algorithm for recursive partitioning multiple populations. We introduce a set of robust criteria that are suitable for different types of response variable. Our criteria need to be robust because of the complexities of the datasets that we have to analyze that usually consist of combinations of clinical studies from different sources, so we do not want a few aberrant observations to dominate the results. The result of our analysis is a list of subsets that represents the characteristics of patients that benefit from drug A more than from drug B. We will illustrate our method by analyzing a group of clinical studies and characterizing the patients who respond better to one drug over competing treatments.

Keywords: ARF; CART; Clinical Trials; Data Mining.

1. Introduction

Mining clinical trial data has become a very useful tool for extracting information that might help understand the performance of drugs or treatments over a population. One important objective that is of interest for all pharmaceutical companies, is to characterize subsets of patients that respond differently than the rest. This information might improve the design of new trials or identify new research targets. For example, are there subsets among the treated group who perform particularly well? Who are the people who don't respond to the drug? Alvir *et al.*[1,2] introduced a tree methodology which we called the ARF (Active Region Finder) methodology that is very useful to answer these kinds of questions directly.

Another set of questions that one might ask are related not so much to the efficacy of a treatment but on how it compares to other treatments or how drug effects compare across different doses. These questions can only be answered indirectly by traditional data mining methods, by adding treatment or dose as a predictor variable and then build a tree or any other data mining model. In cases where the treatment or dose predictor comes out as an important variable in the data mining analysis we might be able to answer the above questions. But it is quite likely that the information is hidden by other factors and might not come up clearly on the analysis.

We proposed a data mining methodology to compare multiple treatments. The objective is to learn about treatment effect differences across the patient populations. One way to do this is to identify regions where treatments effects differ greatly. Each treatment produces its own individual treatment effect function defined over the same patient's space and the objective of our method is to identify subsets of the patient's space where there is a difference. Depending on the specific application we might be interested on a particular treatment being better than the others or where there is a bigger spread among all treatments or in some other measure of difference. The measure of difference is identified by some statistic and it is optimized over regions of the patient's space.

We now take a look at a few examples of how these ideas can be used to answer some of our questions related to clinical methodology. Are there any subgroups of patients or subsets of the patients space where:

(5) A new treatment is more effective than the standard treatment.

(5) Our drug is more effective than all the competitors.

(5) High doses of our drug are more effective than low doses or placebo.

2. Recursive partitioning and active region finder methodology

Recursive partition (RP) algorithms such as CART, C4.5 and FIRM[3,4] have been the most basic data mining methodology for many years. The first important reference on classification and regression trees is the book by Breiman, Friedman, Olshen and Stone (1983).[4] However, the complexity of the clinical trial databases with potentially thousands of patients and hundreds of features makes it very hard to apply standard data mining methodology. Standard RP methods produce very large trees that explain all the variation of the response across the entire dataset, and are tedious and difficult to interpret. For this reason[1,2] introduced the ARF (Active Region Finder) methodology that provides simpler an more focused tree sketches that are able to extract the important subgroups of the data.

An example of this is shown in Fig. 1. The graph is a scatter plot of response to treatment Vs initial pain measured on a standard rating scale for a group of individuals who have been administered a placebo. The response to treatment variable takes values 0 for nonresponders and 1 for responders. The pain scale variable takes values from 0 to 10, where 10 means very high pain and 0 means no pain. The smooth curve across the graph represent the average proportion of responders for a given value of the pain scale variable. The important fact on the graph is that there is an interval (a, b) for which proportion of placebo responders is very high. In the rest of the region the curve is still very nonlinear but the information is not so relevant. When the number of observations is very large, standard recursive partitioning methods will try to explain every detail of the relationship on Fig. 1 and provide a complicated confusing analysis. The interval (a, b) represents the center bucket that was chosen by the ARF algorithm coinciding with the biggest bump in the function. There are other less interesting bumps that are not selected. Hence the difference between ARF and recursive partitioning is that ARF is only interested in the bumps while RP is interested on the whole relationship.

ARF splits are interval splits whereas CART splits are binary. At each node the ARF algorithm will split the data into three buckets defined by two cuts (a, b) on the range of a continuous predictor variable X. The main bucket is the central bucket and it is defined by the values of X that fall inside the interval (a, b), namely $a < X < b$. The left bucket is defined by $X < a$ and right bucket is defined by $X > b$. For binary or categorical predictors the node is split into two buckets. CART splits are always produce two buckets, the left bucket $X < a$ and the right bucket $X > a$.

The criteria that is used in ARF is very simple since we only try to maximize the response mean over all possible subgroups. If a node has a proportion π of 1's then we find the subinterval that maximizes the standard z statistic $z = \frac{(p-\pi)}{\sqrt{\frac{\pi(1-\pi)}{n}}}$. This is very different than the gini index use by CART.[4]

The differences and similarities of both methods have been thoroughly studied[1,2] .

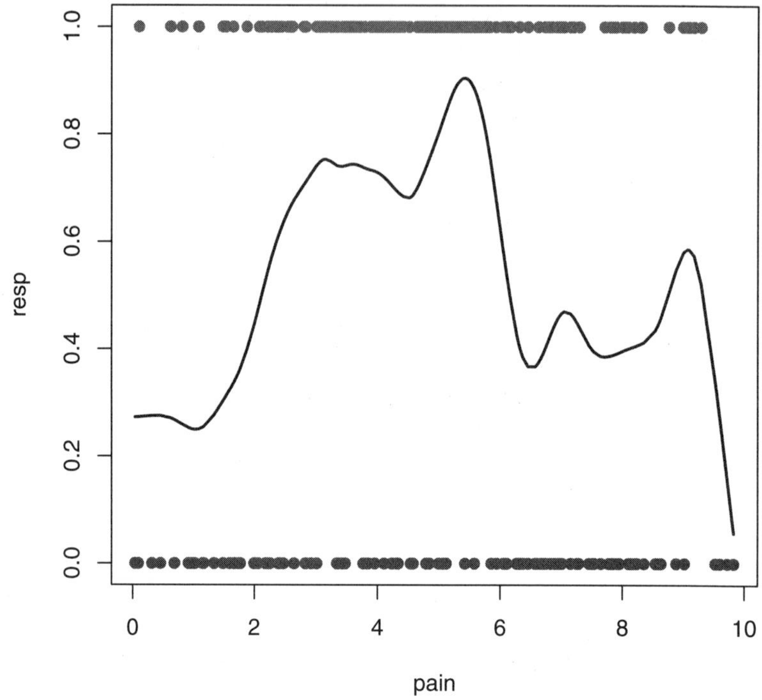

Fig. 1. Response to treatment Vs initial pain measured on a standard rating scale, for a group of individuals who have been administered a placebo.

3. A new data mining paradigm for multiple comparisons

In our case we are dividing the data into multiple groups indexed by a grouping variable, namely I. This index variable I take values over a set of several treatments or doses, $I \in 1, \ldots, k$.

The simplest case is when I is a binary variable differentiating between treatment and control. Therefore $Y(I = 1)$ and $Y(I = 2)$ represent the response observed from a patient in the control or treatment groups respectively. Our objective in this case is to find subgroups of X's for which the $E[Y(I = 2)] - E[Y(I = 1)]$ is maximized.

Our two algorithms on the previous section CART and ARF can be modified to accomplish this task. The partition mechanism will the same as in the previous section but our criteria functions have to be modified in order to accomplish our objective.

For the ARF case, given an interval (a, b) we divided the observations into two groups $G_1 = \{I = 1\}$ and $G_2 = \{I = 2\}$ and let p_1 be the mean of Y over the n_1 observation in G_1 and p_2 the mean of Y over the n_2 observation falling in G_2. Let π be the overall mean of Y over the $n = n_1 + n_2$ observations inside the interval. Then the new statistic becomes $z = \frac{(p_2 - p_1)}{\sqrt{\pi(1-\pi)(\frac{1}{n_1} + \frac{1}{n_2})}}$.

The generalization of the above criteria to multiple treatment groups is straight forward. We developed a software *MPART* in R that incorporates all the above ideas for most reasonable situations. *MPART* covers categorical and numeric responses and predictors, together with two or more treatments.

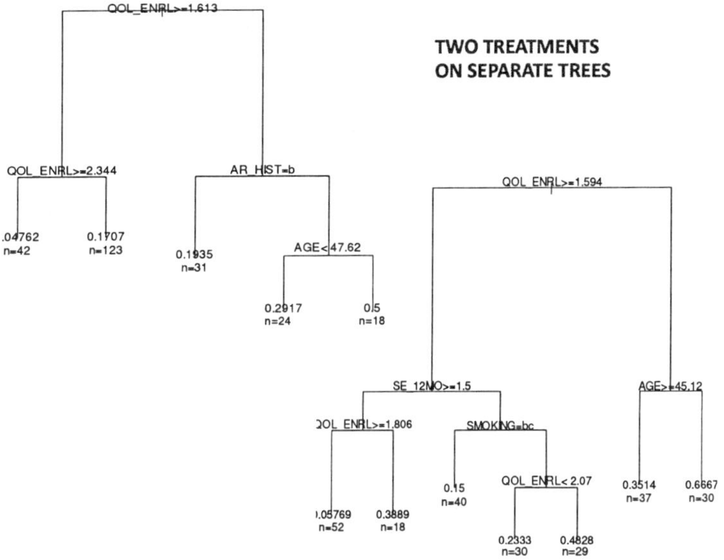

Fig. 2. Clinical study of acute bacterial sinusitis. CART analysis of two treatments on separate tree.

4. Example

We provide an application of our methodology to data from a phase IV, multi-center, open label study that compared azithromycin extended release (AZ-ER) versus amoxicillin/clavulanate potassium (A/C) in subjects with acute bacterial sinusitis (ABS).[5] The study's primary objective was to test whether a single 2.0 g dose of AZ-ER was non-inferior with regard to rate of self-reported symptom resolution at Day 5 compared to 10 days of A/C, 875 mg/125mg q12h. AZ-ER proved superior to A/C on the primary endpoint. More AZ-ER patients (29.7%, 70/236) showed symptom resolution before the end of day 5 than did A/C patients (18.9%, 45/238), a difference of 10.8% (95% CI, 3.1%, 18.4%). Severity of illness at study entry was an important prognostic factor, with milder cases showing greater symptom resolution by day 5. This is evident in Fig. 2 that shows sinusitis-related quality of life at baseline, measured by QOL_ENRL, emerging as the first split in both treatments.

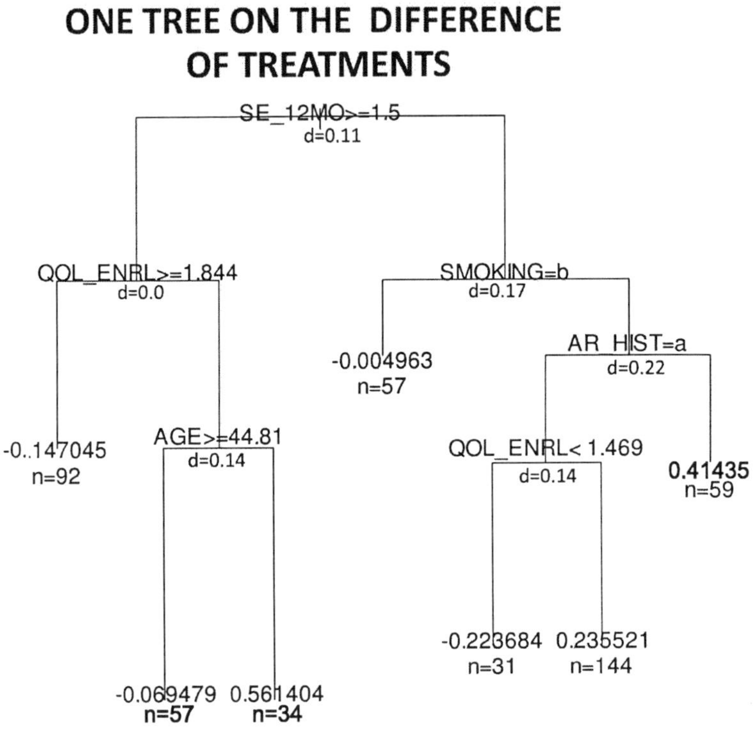

Fig. 3. Clinical study of acute bacterial sinusitis. MPART analysis of the difference of treatment effects.

It was hypothesized that severity might also differentiate response between the two treatments. Application of our methodology to the data indicate that splitting based on the duration of symptoms prior to treatment (SYMP_DUR) provided maximal differentiation, with the greatest difference in patients with 10-12 days of symptoms. The proportion of .41 in the box for the 64 patients who had 10-12 days of symptoms indicates the difference in response rate between AZ-ER and A/C. Among those with shorter duration of symptoms, ratings of sinusitis related quality of life differentiated among patients, with a difference of .32 favoring AZ-ER in a group of 36 patients who were in the middle group of baseline symptom burden (QOL_ENRL between 1.4 & 1.6). This is shown on Fig. 4.

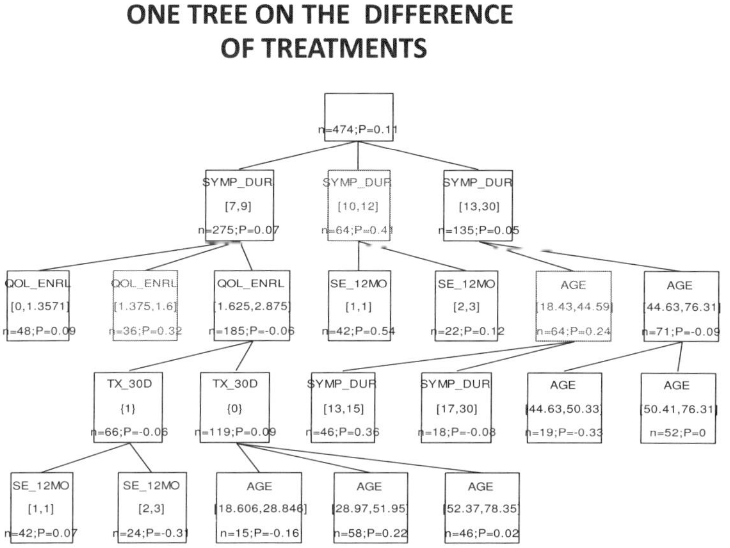

Fig. 4. Clinical study of acute bacterial sinusitis. ARF analysis on the difference of treatment effects.

References

1. E. X. Zhu and I. Davidson (eds.), *Mining Clinical Trial Data* (New York: Information Science Reference, 2007).
2. D. Amaratunga and J. Cabrera, *Journal of Statistical Planning and Inference* **122**, 23 (2004).
3. T. Hastie, R. Tibshirani and J. Friedman, *The Elements of Statistical Learning: data mining, inference and prediction* (Springer Verlag, 2001).

4. B. L., F. J. H., O. R. A. and S. C. J., *Classification and regression trees.* (Wadsworth International Group, Belmont, CA, 1984).
5. J. F. Piccirillo, B. F. Marple, C. S. Roberts, J. R. Frytak, V. F. Schabert, J. C. Wegner, H. Bhattacharyya and S. P. Sanchez, *Symptom Resolution with Azithromycin Extended Release Versus Amoxicillin Clavulanate in Patients with Acute Sinusitis in a General Practice Physician Environment.* 47th Interscience Conference on Antimicrobial Agents and Chemotherapy Chicago, September 17-20, (2007).

SCATTERING OF WATER WAVES BY FREELY FLOATING SEMI-INFINITE ELASTIC PLATES ON WATER OF FINITE DEPTH

ALOKNATH CHAKRABARTI * and SUBASH CHANDRA MARTHA †

Department of Mathematics
Indian Institute of Science, Bangalore 560012, India
* *Tel: +9180 2293 2711, Fax: +9180 2360 0146, Email: alok@math.iisc.ernet.in*
† *Email: scmartha@gmail.com*

A class of mixed boundary value problems (BVPs) arising in the study of scattering of surface water waves by the edges of floating structures comprising of elastic plates, with or without cracks, is examined for their solutions. It is observed that the simplest possible method of solution of such BVPs is the one that involves solution of an over-determined system of Linear Algebraic Equations. Such over-determined systems of equations are best solved by the method of least squares. Numerical results for useful practical quantities such as the "reflection" and "transmission" coefficients are obtained for one of the problems considered here.

Keywords: Flexural gravity waves; Floating elastic plates; Cracks; Scattering.

1. Introduction

Problems of scattering of surface water waves in the two-dimensional linearised theory have created varieties of challenges to applied mathematicians (see Evans and Linton,[5] Stoker,[12] Ursell,[15] Weitz and Keller,[16] amongst others), willing to handle a class of mixed boundary value problems for the two-dimensional Laplace's equation under different types of mixed boundary conditions occurring in the modeling of realistic physical situations applicable to ocean engineering sciences.

Problems of scattering of surface water waves involving floating ice-covers modelled as thin elastic plates have been of major concern for a long time and several papers have been published in this direction (See Blamforth and Craster,[1] Chakrabarti,[3] Marchenko,[8,9] Sahoo, Yip & Chwang,[11] Square and Dixon,[13] Teng *et al.*[14] , and many others, as referred in the paper of Evans and Porter[6]), where varieties of mathematical techniques have been utilized and developed further, which involve the theories of Green's functions, Integral Transforms, the Wiener-Hopf technique, Singular Integral Equations, Eigenfunction expansions and so on (See Evans and

Linton,[5] Stoker,[12] Ursell,[15] Weitz and Keller,[16] also).

In the present work, we have considered the problem of scattering of water waves involving an ocean of finite depth having a rigid bed, whereas on the upper surface of the ocean different boundary conditions are met with on different sides of the lines of discontinuity, constituting the edges of thin elastic plates floating on the surface. Such discontinuous surface boundary conditions are realized in practical problems in which the top surface of the fluid is composed of different distributions of ice particles (see Evans,[4] Gabov *et al.*,[7] Weitz and Keller[16]). A boundary value problem of similar nature, involving an ocean of infinite depth, was handled some time back by Chakrabarti,[2] by the use of singular integral equations of Carleman type.

Here we have adopted eigenfunction expansion methods to handle more general scattering problems involving a single plate, a pair of plates and series of plates with narrow cracks, floating on water of finite depth and have shown that the presently considered general problems can be solved completely by using the method by of least-squares. The solutions obtained here are expected to be useful in the studies of Very Large Floating Systems (VLFS) which involve thin elastic plates.

In the next two sections the methods are explained to determine the solutions of the various boundary value problems under consideration.

2. Scattering of Water Waves by a Single Plate and a Pair of Plates with a Gap

We assume that water in the ocean of finite depth under consideration in the present paper is an ideal and incompressible fluid, the ocean bed is rigid and flat and that there exists a single thin semi-infinite elastic plate or a pair of plates with gap floating on the surface of the fluid. A plane incident wave of small amplitude propagates normally to the edge(s) of the floating plate(s) and we wish to determine the reflected and the transmitted waves after the incident wave hits the edge(s).

2.1. *Formulation of the problem*

We use Cartesian coordinates (x, y), with the line $y = 0$ representing the top surface and the line $y = -h$ the rigid bed. In the case of a single floating plate, the plate occupies the region $y = 0, x > 0$, where as in the case of a pair of plates, these occupy the regions $-\infty < x < -a$, $+a < x < +\infty$ on $y = 0$ whereas the the gap $-a < x < +a$ is assumed to be free to the upper atmosphere.

Under the assumption that the irrotational motion of the fluid is simple harmonic in time with angular frequency ω, we can write the independent velocity potentials for these problems as $\Phi(x,y,t) = Re\{\varphi(x,y)e^{-i\omega t}\}$. Then assuming linear theory, the velocity potential function $\varphi(x,y)$ satisfies the Laplace eq.

$$\frac{\partial^2\varphi}{\partial x^2} + \frac{\partial^2\varphi}{\partial y^2} = 0 \quad \text{in } -\infty < x < \infty, \quad -h \le y \le 0, \tag{1}$$

with the boundary condition

$$\frac{\partial\varphi}{\partial y} = 0 \qquad \text{on} \quad y = -h, -\infty < x < \infty, \tag{2}$$

and the other boundary conditions as described below.

Problem-1.

$$\frac{\partial\varphi}{\partial y} - K\varphi = 0 \qquad \text{on} \quad y = 0, \ -\infty < x < 0, \tag{3}$$

$$\left(K'\frac{\partial^5\varphi}{\partial x^4\partial y} + W\frac{\partial\varphi}{\partial y}\right) - \varphi = 0 \quad \text{on} \quad y = 0, x > 0, \tag{4}$$

$$\frac{\partial^3\varphi}{\partial x^2\partial y} \to 0, \quad \frac{\partial^4\varphi}{\partial x^3\partial y} \to 0 \quad \text{as } x \to 0^-, \ y \to 0^-, \tag{5}$$

$$\varphi(x,y) \to \begin{cases} \left(e^{ik_0x} + Re^{-ik_0x}\right)\dfrac{\cosh k_0(h+y)}{\cosh k_0h}, & x \to -\infty, \\ T\dfrac{\cosh p_0(h+y)}{\cosh p_0h}e^{ip_0x}, & x \to +\infty, \end{cases} \tag{6}$$

Problem-2.

$$\left[K_1'\frac{\partial^4}{\partial x^4} + W_1\right]\frac{\partial\varphi}{\partial y} - \varphi = 0, \text{ on } y = 0, \ -\infty < x < -a, \tag{7}$$

$$\frac{\partial\varphi}{\partial y} - K\varphi = 0, \qquad \text{on } y = 0, \ -a < x < a, \tag{8}$$

$$\left[K_2'\frac{\partial^4}{\partial x^4} + W_2\right]\frac{\partial\varphi}{\partial y} - \varphi = 0, \text{ on } y = 0, \ a < x < \infty, \tag{9}$$

$$\frac{\partial^2}{\partial x^2}\left(\frac{\partial\varphi}{\partial y}\right) = 0, \ \frac{\partial^3}{\partial x^3}\left(\frac{\partial\varphi}{\partial y}\right) = 0, \text{ at } y \to 0, x \to -a^+ \text{ and } y \to 0, x \to +a^-, \tag{10}$$

$$\varphi(x,y) \sim \begin{cases} \left(e^{ip_0^{(1)}x} + Re^{-ip_0^{(1)}x}\right)\dfrac{\cosh p_0^{(1)}(h+y)}{\cosh p_0^{(1)}h}, & \text{for } -\infty < x < -a, \\ \left(R_1e^{-ik_0x} + T_1e^{ik_0x}\right)\dfrac{\cosh k_0(h+y)}{\cosh k_0h}, & \text{for } -a < x < +a, \\ T_2e^{ip_0^{(2)}x}\dfrac{\cosh p_0^{(2)}(h+y)}{\cosh p_0^{(2)}h}, & \text{for } a < x < +\infty, \end{cases} \tag{11}$$

where R, R, R_1 and T, T_1, T_2 are the unknown complex constants, which are related with the reflection and transmission coefficients respectively, $K = \omega^2/g$, g the acceleration due to gravity, $K' = EI/\rho\omega^2$, $K_1' = \frac{E_1 I_1}{\rho\omega^2}$, $K_2' = \frac{E_2 I_2}{\rho\omega^2}$, $W = (\rho g - m_s\omega^2)/\rho\omega^2$, $W_1 = \frac{\rho g - m_{s_1}\omega^2}{\rho\omega^2}$, $W_2 = \frac{\rho g - m_{s_2}\omega^2}{\rho\omega^2}$, ρ the fluid density, E, E_1, E_2 the effective Young's moduli of the floating elastic plates, I, I_1, I_2 the inertial moments of a unit length of each of the elastic plates, and m_s, m_{s_1}, m_{s_2} the surface density of the plates, k_0, p_0, $p_0^{(1)}$, and $p_0^{(2)}$ are positive numbers satisfying, respectively, the transcendental equations

$$K - k_0 \tanh k_0 h = 0, \tag{12a}$$

$$(K' p_0^4 + W) p_0 \tanh p_0 h - 1 = 0, \tag{12b}$$

$$(K_1' p_0^{(1)^4} + W_1) p_0^{(1)} \tanh p_0^{(1)} h - 1 = 0, \tag{12c}$$

$$(K_2' p_0^{(2)^4} + W_2) p_0^{(2)} \tanh p_0^{(2)} h - 1 = 0. \tag{12d}$$

Note that the equation $K + k \tan kh = 0$ has infinite number of real roots of the type $k = \pm k_n, k_n > 0, (n = 1, 2, \ldots)$ and the equation $(K'p^4 + W)p \tan ph + 1 = 0$ has four complex roots of the type, $\pm(\alpha \pm i\beta)$ with $\alpha, \beta > 0$ and has infinite number of real roots of the type $p = \pm p_n$, $p_n > 0, (n = 1, 2, \ldots)$.

2.1. *Method of solution*

Problem-1. We start with the following series representation of the potential function φ (for $-h < y < 0$):

$$\varphi = \left(e^{ik_0 x} + Re^{-ik_0 x}\right) \psi_0(y) + \sum_{m=1}^{\infty} A_m \psi_m(y) e^{k_m x}, \text{ for } x < 0, \tag{13a}$$

and

$$\varphi = T f_0(y) e^{ip_0 x} + \sum_{\substack{n=-2 \\ n \neq 0}}^{\infty} B_n f_n(y) e^{-p_n x}, \text{ for } x > 0, \tag{13b}$$

where A_m's and B_n's are unknown constants to be determined and

$$\psi_m(y) = \begin{cases} \dfrac{\cosh k_0(h+y)}{\cosh k_0 h}, & m = 0, \\[2ex] \dfrac{\cos k_m(h+y)}{\cos k_m h}, & m = 1, 2, \ldots, \end{cases} \tag{14}$$

$$f_n(y) = \begin{cases} \dfrac{\cosh p_0(h+y)}{\cosh p_0 h}, \; n = 0, \\ \\ \dfrac{\cos p_n(h+y)}{\cos p_n h}, \quad n = -2, -1, 1, 2, \dots \quad . \end{cases} \tag{15}$$

Here $p_{-2} = \alpha + i\beta$ and $p_{-1} = \alpha - i\beta$, with $\alpha, \beta > 0$.
We use the conditions of continuity of φ and $\partial\varphi/\partial x$ across $x = 0$ and obtain the following two relations (for $-h < y < 0$):

$$(1+R)\psi_0(y) + \sum_{m=1}^{\infty} A_m \psi_m(y) = T f_0(y) + \sum_{\substack{n=-2 \\ n \neq 0}}^{\infty} B_n f_n(y), \tag{16a}$$

$$ik_0(1-R)\psi_0(y) + \sum_{m=1}^{\infty} A_m k_m \psi_m(y) = T i p_0 f_0(y) - \sum_{\substack{n=-2 \\ n \neq 0}}^{\infty} B_n p_n f_n(y), \tag{16b}$$

Equations (16a) and (16b) represent an over-determined system of linear algebraic equations involving the unknown constants R, T, A_m's and B_n's for which a least-squares solution can be determined by utilizing the following two different ideas:

Idea-I: If we assume that the right sides of Eqs. (16a) and (16b) are known we can obtain a least-squares solution of these equations by using the linearly independent set of functions $\psi_m(y)(m = 0, 1, 2, \dots)$ along with the inner product as defined by

$$\langle f, g \rangle = \int_{-h}^{0} f(y)g(y)dy. \tag{17}$$

We then obtain (assuming $\infty = N$ and m stops at N) a system of $(2N+2)$ number of normal equations involving totally, $(2N+4)$ unknowns, R, T, B_n's $(n = -2, -1, 1, 2, \dots, N)$ and A_m's $(m = 1, 2, \dots, N)$ for the least-squares solution. So we need two more equations which will be supplied by the two edge conditions and hence, we can determine the complete solution of the problem by using a standard method of linear algebra.

Idea-II: If, instead, we assume that the right side is known, and use the set of functions $\{f_n(y)\}(n = -2, -1, 0, 1, 2, \dots)$, we can determine a least-squares solution for the system (16a) and (16b).

Problem-2. For this problem we express the solution φ as (for $-h < y < 0$):

$$\varphi(x,y) = \left(e^{ip_0^{(1)}x} + Re^{-ip_0^{(1)}x}\right) f_0^{(1)}(y) + \sum_{\substack{n=-2\\n\neq 0}}^{\infty} B_n f_n^{(1)}(y) e^{p_n^{(1)}x},$$

$$\text{for } -\infty < x < -a, \quad \text{(18a)}$$

$$\varphi(x,y) = \left(R_1 e^{-ik_0x} + T_1 e^{ik_0x}\right) \psi_0(y) + \sum_{m=1}^{\infty} \left[C_m \sinh k_m(x-a) + \right.$$

$$\left. D_m \sinh k_m(x+a)\right] \times \psi_m(y), \text{ for } -a < x < +a, \quad \text{(18b)}$$

$$\varphi(x,y) = T_2 f_0^{(2)}(y) e^{ip_0^{(2)}x} + \sum_{\substack{n=-2\\n\neq 0}}^{\infty} A_n f_n^{(2)}(y) e^{-p_n^{(2)}x}, \text{ for } a < x < \infty, \text{(18c)}$$

where A_n, B_n $(n = -2, -1, 1, 2, \ldots)$ and $C_m, D_m (m = 1, 2, \ldots)$ are unknown constants which can be determined along with R, R_1, T_1 and T_2 by using the continuity conditions of φ and φ_x across $x = 0$, $x = +a$ and $x = -a$ as well as the edge conditions, similar to the ones used in problem 1, with (for $i = 1, 2$)

$$f_n^{(i)}(y) = \begin{cases} \dfrac{\cosh p_0^{(i)}(h+y)}{\cosh p_0^{(i)}h}, & n = 0, \\[2ex] \dfrac{\cos p_n^{(i)}(h+y)}{\cos p_n^{(i)}h}, & n = -2, -1, 1, 2, \ldots \end{cases} \quad (19)$$

We obtain two different systems of over-determined equations in the case of the present problem (the details are omitted), involving the above unknown constants, for which we can determine least-squares solutions by using methods, similar to the ones used in problem-1.

3. Problem-3. Scattering of Flexural Gravity Waves by a Series of Narrow Cracks in a Floating Elastic Plate

This problem arises in studying (see Porter and Evans[10]) the scattering of obliquely incident flexural gravity waves involving water of finite depth h, on the surface $y = 0$ of which there lies an infinite elastic plate with known parameters β, δ and ν, having narrow straight-line cracks along the lines $x = 0, a_1, a_2, \ldots, a_n$. We adopt here an approach which is slightly different from that of Porter and Evans[10] . Leaving aside details, the equations and conditions for the problem are given by:

$$\frac{\partial^2 \varphi}{\partial x^2} + \frac{\partial^2 \varphi}{\partial y^2} - l^2 \varphi = 0, \text{ in the region } -h \leq y \leq 0, \; -\infty < x < \infty, \quad (20)$$

$$\frac{\partial\varphi}{\partial y} = 0, \text{on} \quad y = -h, -\infty < x < \infty, \tag{21}$$

$$\left[\beta\left(\frac{\partial^2}{\partial x^2} - l^2\right)^2 + 1 - \delta\right]\frac{\partial\varphi}{\partial y} - \varphi = 0, \text{on} \quad y = 0,\ x \in (-\infty, 0) \cup (0, a_1)$$
$$\cup(a_1, a_2) \ldots \cup (a_{p-1}, a_p) \cup (a_p, \infty), \tag{22}$$

$$\mathfrak{B}\varphi = \left[\left(\frac{\partial^2}{\partial x^2} - \nu l^2\right)\frac{\partial\varphi}{\partial y}\right]_{y=0} \to 0, \text{ as } x \to (0^{\pm}, a_1^{\pm}, \ldots, a_p^{\pm}), \tag{23}$$

$$\mathscr{S}\varphi = \left[\frac{\partial}{\partial x}\left(\frac{\partial^2}{\partial x^2} - \nu_1 l^2\right)\frac{\partial\varphi}{\partial y}\right]_{y=0} \to 0, \text{ as } x \to (0^{\pm}, a_1^{\pm}, \ldots, a_p^{\pm}), \tag{24}$$

$$\varphi(x, y) \to \begin{cases} \varphi_0(x, y) + R\varphi_0(-x, y), & x \to -\infty, \\ T\varphi_0(x, y), & x \to +\infty, \end{cases} \tag{25}$$

where

$$\varphi_0(x, y) = e^{ik_0 x}\cosh\gamma_0(h + y), \tag{26}$$

$\beta, \delta, l, \nu, a_1, a_2, \ldots, a_p$ are all known positive constants, $\nu_1 = 2 - \nu$, $\gamma_0 > l$ and γ_ns (with $\gamma_n^2 = k_n^2 + l^2$, $(n = -2, -1, 0, 1, 2, \ldots)$) are the roots of the eq.

$$K(k_n) \equiv (\beta\gamma_n^4 + 1 - \delta)\gamma_n\tanh(\gamma_n h) - 1 = 0. \tag{27}$$

3.1. *The method of solution*

The solution of the boundary value problem as posed by the relations (20)-(22), can be expressed in the following forms

$$\varphi(x, y) = \begin{cases} \varphi_0(x, y) + \sum\limits_{n=-2}^{\infty} B_n^{(0)} e^{-ik_n x} Y_n(y), & x < 0 \\ \varphi_0(x, y) + \sum\limits_{n=-2}^{\infty} \left[A_n^{(1)} e^{ik_n x} + B_n^{(1)} e^{-ik_n x}\right] Y_n(y), & 0 < x < a_1 \\ \cdots & \\ \varphi_0(x, y) + \sum\limits_{n=-2}^{\infty} \left[A_n^{(p)} e^{ik_n x} + B_n^{(p)} e^{-ik_n x}\right] Y_n(y), & a_{p-1} < x < a_p \\ \varphi_0(x, y) + \sum\limits_{n=-2}^{\infty} A_n^{(p+1)} e^{ik_n x} Y_n(y), & x > a_p, \end{cases} \tag{28}$$

where $Y_n(y) = \cosh\gamma_n(h + y), (-h < y < 0), (n = -2, -1, 0, 1, 2, \ldots)$ and the constants $A_n^{(r)}(r = 1, 2, \ldots, p + 1)$ and $B_n^{(r)}(r = 0, 1, 2, \ldots, p)$,

$(n = -2, -1, 0, 1, 2, \ldots)$ can be determined by utilizing the conditions of continuity of φ and $\varphi_x \left(\equiv \frac{\partial \varphi}{\partial x}\right)$ across the lines $x = 0, a_1, a_2, \ldots, a_p$, and the edge conditions (23) and (24), as described above.

The continuity of φ and φ_x across the lines $x = 0, a_1, a_2, \ldots, a_p$ give rise to over-determined systems of $(2p+2)$ linear equations for the unknown constants $A_n^{(r)}(r = 1, 2, \ldots, p+1)$ and $B_n^{(r)}(r = 0, 1, 2, \ldots, p)$, $(n = -2, -1, 0, 1, 2, \ldots)$.

We have determined the unknown constants by utilizing the following identity (see Appendix):

$$\sum_{n=-2}^{\infty} \frac{(\gamma_n^2 + \theta)}{C_n} Y_n'(0) Y_n(y) = 0, \quad (-h < y < 0) \tag{29}$$

where θ is any constant and

$$C_n = \frac{1}{2}\left[h + (5\beta\gamma_n^4 + 1 - \delta)\left\{\frac{Y_n'(0)}{\gamma_n}\right\}^2\right] \tag{30}$$

and have obtained the following relations:

$$\left[\left\{B_n^{(0)} - B_n^{(1)}\right\} - A_n^{(1)}\right] = P_0(\gamma_n^2 + \theta_0)C_n^{-1}Y_n'(0), \tag{31a}$$

....

$$\text{and } \left[\left\{A_n^{(p)} - A_n^{(p+1)}\right\} e^{ik_n a_p} + B_n^{(p)} e^{-ik_n a_p}\right] = P_p(\gamma_n^2 + \theta_p)C_n^{-1}Y_n'(0), \tag{31b}$$

and

$$k_n\left[A_n^{(1)} - \left\{B_n^{(1)} - B_n^{(0)}\right\}\right] = Q_0(\gamma_n^2 + \lambda_0)C_n^{-1}Y_n'(0), \tag{32a}$$

....

$$\text{and } k_n\left[\left\{A_n^{(p+1)} - A_n^{(p)}\right\} e^{ik_n a_p} + B_n^{(p)} e^{-ik_n a_p}\right] = Q_p(\gamma_n^2 + \lambda_p)C_n^{-1}Y_n'(0), \tag{32b}$$

where $P_r's, Q_r's, \theta_r's$ and $\lambda_r's$, $(r = 0, 1, 2, \ldots, p)$ are $4(p+1)$ unknown constants which can be determined fully by using the edge conditions (23) and (24) along with the relations (28). We have also used an extremely useful identity as given by

$$\sum_{n=-2}^{\infty} \frac{(k_n^2 + \nu l^2)(k_n^2 + \nu_1 l^2)\{Y_n'(0)\}^2}{C_n} = 0. \tag{33}$$

(see Evans and Porter[6]) to simplify the final results.

Special cases

We consider below some special cases of this problem and present analytical as well as numerical results for the reflection and transmission coefficients.

Problem involving one crack:

We find that in the case of only one crack, situated at $x = 0$, i.e., for $p = 0$, we obtain the reflection and transmission coefficients R and T, respectively, in the regions $x < 0$ and $x > 0$, as given by

$$R = B_0^{(0)} = \frac{1}{2}[P_0 U_0 + Q_0 V_0], \tag{34}$$

$$T = 1 + A_0^{(1)} = 1 + \frac{1}{2}[-P_0 U_0 + Q_0 V_0], \tag{35}$$

with P_0 and Q_0 as given by

$$P_0 = \frac{2k_0(k_0^2 + \nu_1 l^2)Y_0'(0)}{\sum_{n=-2}^{\infty} \frac{k_n}{C_n}(k_n^2 + \nu_1 l^2)^2 \{Y_n'(0)\}^2}, \quad Q_0 = \frac{-2(k_0^2 + \nu l^2)Y_0'(0)}{\sum_{n=-2}^{\infty} \frac{k_n^{-1}}{C_n}(k_n^2 + \nu l^2)^2 \{Y_n'(0)\}^2}, \tag{36}$$

and U_0 and V_0 are given by Eq. (40) (see later) for $n = 0$. These results determining the complete solution of the problem involving a single crack are in full agreement with the results obtained by Evans and Porter[6] and we find that these solutions are determined without breaking the problem into symmetric and asymmetric parts.

Problem involving a pair of cracks:

In the case of the problem involving a pair ($p = 1$) of cracks, we find the four important constants, R_1, R_2, T_1 and T_2, in the regions $x < 0$, $0 < x < a$ and $x > a$, in the following forms:

$$\left.\begin{aligned} R_1 &= B_0^{(0)} = \tfrac{1}{2}\left[P_0 U_0 + Q_0 V_0 + (P_1 U_0 + Q_1 V_0)e^{ik_0 a}\right], \\ R_2 &= B_0^{(1)} = \tfrac{1}{2}(P_1 U_0 + Q_1 V_0)e^{ik_0 a}, \\ T_1 &= 1 + A_0^{(1)} = 1 - \tfrac{1}{2}[P_0 U_0 - Q_0 V_0], \\ T_2 &= 1 + A_0^{(2)} = 1 + \tfrac{1}{2}\left[(Q_1 V_0 - P_1 U_0)e^{-ik_0 a} - (P_0 U_0 - Q_0 V_0)\right], \end{aligned}\right\} \tag{37}$$

where the new constants P_0, P_1, Q_0, Q_1 can be determined by solving the system of linear algebraic equations, as given by

$$A\vec{q} = \vec{b}, \tag{38}$$

with the matrix $A = [a_{ij}]$ $(i, j = 1, 2, 3, 4)$ and the vector $\vec{b} = [b_1, b_2, b_3, b_4]^T$, being given by the following relations, with $\vec{q} =$

$[P_0, P_1, Q_0, Q_1]^T$:

$$
\begin{aligned}
&a_{11} = 0 = a_{22} = a_{33} = a_{44}, a_{12} = -\sum_{n=-2}^{\infty} (k_n^2 + \nu l^2) V_n Y_n'(0),\\
&a_{13} = \sum_{n=-2}^{\infty} -(k_n^2 + \nu l^2) U_n e^{ik_n a} Y_n'(0), a_{14} = -\sum_{n=-2}^{\infty} (k_n^2 + \nu l^2) V_n e^{ik_n a} Y_n'(0),\\
&a_{21} = \sum_{n=-2}^{\infty} k_n (k_n^2 + \nu_1 l^2) U_n Y_n'(0), a_{23} = \sum_{n=-2}^{\infty} k_n (k_n^2 + \nu_1 l^2) U_n e^{ik_n a} Y_n'(0),\\
&a_{24} = \sum_{n=-2}^{\infty} k_n (k_n^2 + \nu_1 l^2) V_n e^{ik_n a} Y_n'(0),\\
&a_{31} = -a_{13},\ a_{32} = a_{14},\ a_{34} = a_{12}, a_{41} = a_{23},\ a_{42} = -a_{24},\ a_{43} = a_{21},\\
&b_1 = 2(k_0^2 + \nu l^2) Y_0'(0), \quad b_2 = 2k_0 (k_0^2 + \nu_1 l^2) Y_0'(0),\\
&b_3 = 2(k_0^2 + \nu l^2) e^{ik_0 a} Y_0'(0), \quad b_4 = 2k_0 (k_0^2 + \nu_1 l^2) e^{ik_0 a} Y_0'(0),
\end{aligned} \tag{39}
$$

with

$$
U_n = \frac{\{\gamma_n^2 + (1-\nu) l^2\} Y_n'(0)}{C_n}, \quad V_n = \frac{\{\gamma_n^2 + (1-\nu_1) l^2\} Y_n'(0)}{k_n C_n}. \tag{40}
$$

3.2. *Numerical results*

In order to obtain various numerical results, we follow the same procedure as described in Evans and Porter.[6] We present the variation of $|R|$ and $|T|$ with wavelength $\bar{\lambda}/\bar{d}$ in the problem involving a single crack and $|R_1|$ and $|T_2|$ with wavelength $\bar{\lambda}/\bar{d}$ in the problem involving a pair of cracks for $\beta' = 45536$ (considering $\bar{d} = 1$), $\bar{h}/\bar{d} = 80$. In both the problems, from tables 1, 2 we find that the reflection and transmission coefficients satisfy the energy-balance relations: (i)$|R|^2 + |T|^2 = 1$ in the case of problem involving a single crack and (ii) $|R_1|^2 + |T_2|^2 = 1$ in the case of problem involving a pair of cracks.

Acknowledgement

The material of the paper was presented by Professor A. Chakrabarti, at the FACM 2008 conference, as a talk in the Minisymposia. AC thanks the organizing committee of the conference for the invitation. AC also thanks the National Board for Higher Mathematics (NBHM), India for travel support and the University Grants Commission (UGC), New Delhi, India for offering an Emeritus Fellowship. SCM is grateful to the NBHM, India for providing a Post Doctoral Fellowship.

Table 1. Variation of $|R|$ and $|T|$ with wavelength for $\beta' = 45536$ ($\bar{d} = 1$), $\bar{h}/\bar{d} = 80$ and $\theta = 0$ (normal incidence) in the problem involving a single crack

$\bar{\lambda}/\bar{d}$	$\|R\|$	$\|T\|$	$\|R\|^2 + \|T\|^2$
10	0.7321	0.6812	1.0000
11	0.6599	0.7514	1.0000
12	0.6956	0.7185	1.0000
13	0.5712	0.8208	1.0000
14	0.5790	0.8154	1.0000
15	0.7050	0.7093	1.0000

Table 2. Variation of $|R_1|$ and $|T_2|$ with wavelength for $\bar{d} = 1, \beta' = 45536$, $\bar{h}/\bar{d} = 80$ and $\theta = 0$ (normal incidence) in the problem involving a pair of cracks

$\bar{\lambda}/\bar{d}$	$\|R_1\|$	$\|R_2\|$	$\|T_1\|$	$\|T_2\|$	$\|R_1\|^2 + \|T_2\|^2$
10	1.7353×10^{-4}	5.6229×10^{-5}	0.4571	1.000	1.000
11	2.2856×10^{-4}	8.7028×10^{-5}	0.5539	1.000	1.000
12	3.1777×10^{-4}	1.1604×10^{-4}	0.5271	1.000	1.000
13	3.5638×10^{-4}	1.5352×10^{-4}	0.5965	1.000	1.000
14	5.8958×10^{-4}	2.1196×10^{-4}	0.5264	1.000	1.000
15	7.3980×10^{-4}	2.7071×10^{-4}	0.5317	1.000	1.000

Appendix A.

Let a_n's be constants to be determined such that

$$\sum_{n=-2}^{\infty} a_n Y_n(y) = 0, \quad -h < y < 0. \tag{A.1}$$

Multiplying both sides of the Eq. (A.1) by $Y_m(y)$ and integrating between $y = -h$ and $y = 0$, we obtain the relation

$$a_m C_m - \beta Y_m'(0) \sum_{n=-2}^{\infty} a_n(\gamma_m^2 + \gamma_n^2) Y_n'(0) = 0, \tag{A.2}$$

after using the result

$$\int_{-h}^{0} Y_m(y) Y_n(y) dy = C_n \delta_{mn} - \beta(\gamma_m^2 + \gamma_n^2) Y_m'(0) Y_n'(0) \tag{A.3}$$

with C_n as defined by the relation (30).

The relation (A.2) helps establishing the fact that

$$a_m C_m = \left(\mu_1 + \mu_2 \gamma_m^2\right) Y_m'(0), \tag{A.4}$$

where μ_1 and μ_2 are two unknown constants.

It is obvious, then, that the relation (A.4) gives the general form of the constants a_m, satisfying the relation (A.1), and a special form of a_m is given by

$$a_m = \frac{(\gamma_m^2 + \theta)}{C_m} Y'_m(0), \tag{A.5}$$

showing that the identity (29) holds good, where θ is any constant.

References

1. N.J. Balmforth and R.V. Craster, Ocean waves and ice sheets, J. Fluid Mech. **395**, 089-124 (1999).
2. A. Chakrabarti, On the solution of the problem of scattering of surface water waves by a sharp discontinuity in the surface boundary conditions, Anziam J. **42**, 277-286 (2000).
3. A. Chakrabarti, On the solution of the problem of scattering of surface-water waves by the edge of an ice-cover, Proc. R. Soc. Lond. A **456**, 1087-1099 (2000).
4. D.V. Evans, The solution of a class of boundary value problems with smoothly varying boundary conditions, Q.J. Mech. Appl. Math., **38**, 521-536 (1994).
5. D.V. Evans and C.M. Linton, On the step approximation for water wave problems, J.Fluid Mech. **278**, 229-249 (1994).
6. D.V. Evans and R. Porter, Wave scattering by narrow cracks in ice sheets floating on water of finite depth, J. Fluid Mech. **484**, 143-165 (2003).
7. S.A. Gabov, A.G. Svesnikov and A.K. Shatov, Dispersion of internal waves by an obstacle floating on the boundary separating two liquids, Prikl. Mat. Mech. (Russian) **53**, 727-730 (1989).
8. A.V. Marcheno, Surface wave diffraction at a crack in sheet ice, Fluid Dyn. **28**, 230-237 (1993).
9. A.V. Marcheno, Flexural gravity wave diffraction at linear irregularities in sheet ice, Fluid Dyn. **32**, 548-560 (1997).
10. R. Porter and D.V. Evans, Scattering of flexural waves by multiple narrow cracks in ice sheets floating on water, Wave Motion **43**, 425-443 (2006).
11. T. Sahoo, T.L. Yip and A.T. Chwang, Scattering of surface waves by a semi-infinite floating elastic plate, Phys. Fluids **13**, 3215-3222 (2001).
12. J.J. Stoker, Water Waves, Wiley Interscience, New York (1957).
13. V.A. Squire and T.W. Dixon, An analytic model for wave propagation across a crack in an ice sheet, Intl. J. Offshore Polar Engng, **10**, 173-176 (2000).
14. B. Teng, L. Cheng, S.X. Liu , F.J. Li, Modified eigenfunction expansion methods for interaction of water waves with a semi-infinite elastic plate. Appl Ocean Res **23**, 357-368 (2001).
15. F. Ursell, The effect of a fixed vertical barrier on surface water waves in deep water, Proc. Cambridge Philos.Soc. **43**, 374-382 (1947).
16. M. Weitz and J.B. Keller, Reflection of water waves from floating ice in water of finite depth, Comm. Pure Appl. Math. **3**, 305-318 (1950).

ON THE HYPERBOLICITY OF TWO-LAYER FLOWS

RICARDO BARROS and WOOYOUNG CHOI

Department of Mathematical Sciences
Center for Applied Mathematics and Statistics
New Jersey Institute of Technology
Newark, NJ 07102-1982, USA

We consider the two-layer shallow water equations in the presence of the top free surface and find explicit conditions for which the system is hyperbolic. It is commonly believed that, analogously to the rigid-lid case, this can only happen for small relative speeds. Using both the root location criteria for a quartic equation and a geometrical approach, it is shown that hyperbolicity is held for not only small, but also large relative speeds.

Keywords: Hyperbolicity; Two-layer flows; Shallow water; Quartic equation.

1. Introduction

We consider a two-layer system composed of two immiscible fluids of different *constant densities* ρ_1 and ρ_2 confined between the upper free surface and the lower rigid boundary in which the flow is governed by the two-layer shallow water equations (e.g., see Baines[1]):

$$\begin{gathered}(h_i)_t + (h_i\,u_i)_x = 0,\\ (u_i)_t + u_i(u_i)_x + g\big(h_1 + h_2 + \delta_{i1}(\rho - 1)h_2\big)_x = 0.\end{gathered} \tag{1}$$

In these equations, u_1 and u_2 are the depth-averaged velocities, h_1 and h_2 are the layer thicknesses, δ_{ij} is the Kronecker delta, g is the gravitational acceleration, and $\rho < 1$ is defined by $\rho = \rho_2/\rho_1$, with 1 and 2 associated with the lower and upper layer, respectively. By defining $U = (h_1, h_2, u_1, u_2)^T$, the quasilinear system can be written in the form $U_t + A\,U_x = 0$, for which the characteristic polynomial $P(\lambda) = \det(A - \lambda I)$ is given by

$$\begin{aligned}P(\lambda) &= \lambda^4 - 2(u_1 + u_2)\lambda^3 + (u_1^2 + 4u_1u_2 + u_2^2 - gh_1 - gh_2)\lambda^2 + 2\,(gh_1u_2\\ &+ gh_2u_1 - u_1u_2^2 - u_2u_1^2)\lambda + u_1^2u_2^2 - gh_1u_2^2 - gh_2u_1^2 + g^2h_1h_2(1 - \rho).\end{aligned} \tag{2}$$

The system (1) is hyperbolic if $P(\lambda) = 0$ admits only real roots. It seems to be widely accepted that this can happen only if the relative speed between the two layers is small (see Refs. 2–7). Approximate expressions for the eigenvalues of A valid in the Boussinesq limit ($\rho \approx 1$) were first obtained by Schijf & Schnfeld[2] while the exact expressions for the eigenvalues were found by Lawrence[4] without relying on the Boussinesq approximation. This result has been recently extended in Ref. 6 with including the effects of topography. It is possible to present the eigenvalues (or characteristic speeds) in two distinct sets corresponding to the *external* and *internal* wave motions, respectively. All these authors seem to agree that the eigenvalues corresponding to the internal wave mode become complex when the *Froude number* F defined by $F = (u_2 - u_1)/\sqrt{gh_1}$ exceeds a critical value. However, this seems to be in contradiction with the result obtained by Ovsyannikov,[8] who showed, by means of a geometrical representation of the characteristics, that the model can still be hyperbolic for large relative speeds. In this paper, by examining carefully the criteria for the quartic equation (2) to have real roots, we validate the result of Ref. 8 that the internal wave speeds can indeed be complex only for a bounded range of Froude numbers.

2. Hyperbolicity of the two-layer shallow water model

2.1. *Root location criteria*

Before proceeding further, we summarize below some essentials of the root distribution for the quartic equation.

2.1.1. *Preliminaries*

Consider the quartic equation

$$f(x) = a_0 x^4 + a_1 x^3 + a_2 x^2 + a_3 x + a_4 = 0, \tag{3}$$

whose coefficients a_i $(i = 0, \cdots, 4)$ are all real with $a_0 > 0$. It is known that the *discriminant* Δ_f of Eq. (3) is given by

$$\Delta_f = {a_0}^6 \prod_{i<k} (x_i - x_j)^2,$$

where x_1, x_2, x_3, and x_4 are the roots of the polynomial f. Since Δ_f is a symmetric polynomial, it can be polynomially expressed in terms of the real coefficients a_i (see page 80 of Ref. 9). The discriminant is a powerful tool that can fully describe the structure of the roots for quadratic and cubic equations. However, the same cannot be achieved for the quartic equation since (i) $\Delta_f > 0$: four distinct real or four distinct complex roots;

(*ii*) $\Delta_f = 0$: at least two equal roots; (*iii*) $\Delta_f < 0$: two distinct real roots and two complex roots.

Several attempts have been made in the past to obtain conditions, in terms of the literal coefficients of a polynomial, concerning a special root distribution (see references therein). Among them, Jury & Mansour[10] presented a series of algorithms involving *characteristic expressions* for a quartic equation, allowing a full characterization of the root distribution in a much more concise form than the one provided by previous approaches. Similar criteria involving only inner determinants were also obtained by Fuller.[11] Following this elegant exposition, when considering the inner determinants Δ_3, Δ_5, Δ_7 [a]:

$$\Delta_7 = \begin{vmatrix} a_0 & a_1 & a_2 & a_3 & a_4 & 0 & 0 \\ 0 & a_0 & a_1 & a_2 & a_3 & a_4 & 0 \\ 0 & 0 & a_0 & a_1 & a_2 & a_3 & a_4 \\ 0 & 0 & 0 & 4a_0 & 3a_1 & 2a_2 & a_3 \\ 0 & 0 & 4a_0 & 3a_1 & 2a_2 & a_3 & 0 \\ 0 & 4a_0 & 3a_1 & 2a_2 & a_3 & 0 & 0 \\ 4a_0 & 3a_1 & 2a_2 & a_3 & 0 & 0 & 0 \end{vmatrix},$$

with Δ_3 and Δ_5 being defined as the determinants of the inner matrices with dimensions 3×3 and 5×5, respectively (as denoted by the two inner squares in the definition of Δ_7), we have the following result (see page 778 of Fuller[11]):

Theorem 2.1. *Equation* (3) *has its roots all real if and only if one of the two following sets of conditions holds:* (*a*) $\Delta_3 > 0$, $\Delta_5 > 0$, $\Delta_7 \geqslant 0$; (*b*) $\Delta_3 \geqslant 0$, $\Delta_5 = 0$, $\Delta_7 = 0$.

2.1.2. *Real characteristic speeds of long waves*

We first rewrite (2) in terms of non-dimensional variables:

$$\Lambda = \frac{\lambda}{\sqrt{gh_1}}, \qquad H = \frac{h_2}{h_1}, \qquad F = \frac{u_2 - u_1}{\sqrt{gh_1}}$$

and assume without loss of generality that $u_1/\sqrt{gh_1} = 1$ (by choosing a moving reference frame such that this condition is met). Then, the characteristic equation $P(\lambda) = 0$ becomes

$$\Lambda^4 - 2(2+F)\,\Lambda^3 + [(1+F)(5+F) - H]\,\Lambda^2 + 2\left[H - (1+F)^2\right]\Lambda - \rho H = 0,$$

[a]Notice that Δ_7 is precisely the determinant of the Sylvester matrix, hence it consists on the discriminant of f, *i.e.*, $\Delta_7 = \Delta_f$.

for which the inner determinants are found as

$$\Delta_3 = 2F^2 + 8(H+1),$$

$$\Delta_5 = 8(H+1)F^4 - 16\left[H^2 - (6+\rho)H + 1\right]F^2 + 8(H+1)\left[(H-1)^2 + 4\rho H)\right].$$

We will show that, in our particular case, the set of conditions stated in Theorem 2.1 reduces to $\Delta_7 \geqslant 0$, as the conditions $\Delta_3 > 0$ and $\Delta_5 > 0$ are automatically satisfied. First, it is clear that $\Delta_3 > 0$. Second, in order to prove that $\Delta_5 > 0$, we look at Δ_5 as a parabola in terms of the variable $y = F^2$. If its discriminant is less than zero, the parabola has no real roots and, therefore, $\Delta_5 > 0$. On the other hand, if it has real roots, it is sufficient to prove that both roots are negative, which assures that Δ_5 has no real roots for the Froude number F. The discriminant of Δ_5 becomes positive only if

$$6(\rho+2)H^2 - \left[36 + (\rho+2)^2\right]H + 6(\rho+2) < 0, \tag{4}$$

which holds for $\frac{\rho+2}{6} < H < \frac{6}{\rho+2}$. For this range of H, however, the coefficient of F^2 in Δ_5 is always positive and so the result follows.

We have shown that, in our particular case, a full description of the root distribution can be achieved by means of its discriminant. Therefore, from Theorem 2.1, we can conclude that $P(\lambda) = 0$ has four real solutions for $\Delta_7 \geqslant 0$, while it has two complex and two real solutions for $\Delta_7 < 0$.

Straightforward calculations reveal that the discriminant Δ_7 depends only on the variable $y = F^2$ and the physical parameters ρ and H:

$$\Delta_7 = 16H\,Q(y), \tag{5}$$

with $Q(y)$ defined by

$$\begin{aligned} Q(y) = y^4 &+ (H+1)(\rho-4)\,y^3 - [3(\rho-2) - (4 - 26\rho + \rho^2)H \\ &+ 3(\rho-2)H^2]y^2 + (H+1)[3\rho - 4 + (8 + 10\rho - 20\rho^2)H \\ &+ (3\rho-4)H^2]y + (1-\rho)\left((H-1)^2 + 4\rho H\right)^2. \end{aligned}$$

As shown in Fig. 1, for any prescribed values for the physical parameters, the polynomial $Q(y)$ has two positive real roots $(F^-_{\text{crit}})^2$ and $(F^+_{\text{crit}})^2$, and the condition $\Delta_7 \geqslant 0$ is satisfied for $0 \leqslant F^2 \leqslant (F^-_{\text{crit}})^2$ or $F^2 \geqslant (F^+_{\text{crit}})^2$. The first inequality for the Froude number implies that the system (1) is hyperbolic for small relative speeds between the two layers, as noted by several authors. However, the figure shows clearly a new range of Froude numbers characterized by large relative speeds, for which the flow is hyperbolic (*cf.* Ovsyannikov[8]).

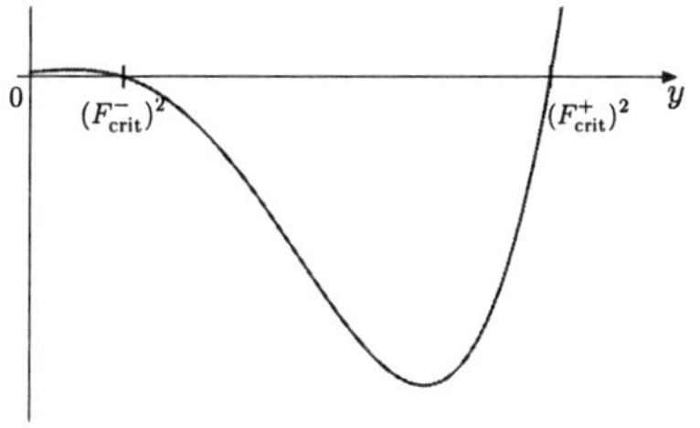

Fig. 1. A sketch of the behavior of the polynomial $Q(y)$

2.1.3. *Comparison with the characteristic speeds in Lawrence (1990)*

We find in the Appendix of Ref. 4 an exact derivation of the characteristic speeds of long waves, based on the Descartes-Euler solution expressed by means of the solutions of the cubic resolvent (A.6). The discriminant of this cubic resolvent given by (A.8) is denoted by D in Ref. 4 whose sign is determined by the quantity δ (with the opposite sign of D) defined by

$$\delta = \beta + (1 - F_\Delta^2) \sum_{n=0}^{3} b_n \epsilon^n,$$

or, alternatively, in our notation,

$$\delta = \frac{4}{(1+H^2)^4(1-\rho)} Q(y). \tag{6}$$

The characteristic speeds are all real provided that $D \leqslant 0$ and the roots of the cubic resolvent are positive. In his work, Lawrence[4] stated that *the requirement that $D \leqslant 0$ for the solutions to be real restricts the value of the Froude number F_Δ^2 to be less than or equal to a critical value $(F_\Delta^2)_{\rm crit}$*, which contradicts our finding. We know from Eq. (6) that, since D has the opposite sign of δ, $D \leqslant 0$ is equivalent to $Q(y) \geqslant 0$, which implies the statement on hyperbolicity in Ref. 4 is inaccurate.

2.2. *A geometrical approach*

The results presented so far will now become more clear. Using the approach proposed in Ref. 8, we will be able to provide a geometrical interpretation for the roots of Δ_7 in Eq. (5). The key step is to rewrite the characteristic polynomial in a simpler form:

$$P(\lambda) = \big((u_1 - \lambda)^2 - gh_1\big)\big((u_2 - \lambda)^2 - gh_2\big) - g^2 \rho h_1 h_2.$$

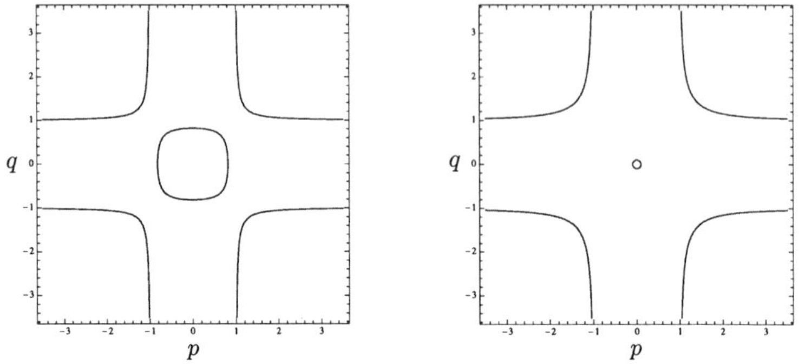

Fig. 2. Plots of the curve (8) for different physical parameters: $\rho = 1/3$ (*left-hand side*) and $\rho = 99/100$ (*right-hand side*).

This form of presenting Eq. (2) is not new, but, as shown in Ref. 8, allows us to better understand the structure of the roots for the characteristic equation. If we define

$$\lambda - u_1 = q\sqrt{gh_1}, \quad \lambda - u_2 = p\sqrt{gh_2}, \tag{7}$$

the characteristic equation yields

$$(p^2 - 1)(q^2 - 1) = \rho. \tag{8}$$

On the (p, q) plane, Eq. (8) describes a fourth-order curve having four axes of symmetry where we can distinguish an inner region (in the interior of the unit square centered at the origin) and an outer region, as shown in Fig. 2. The limit cases correspond to assigning to ρ the values 0 and 1. In the first case ($\rho = 0$), Eq. (8) reduces to the lines $|p| = 1$ and $|q| = 1$. In the second case ($\rho = 1$), the inner region confines to one single point, the origin, confirming its tendency to shrink as the values of ρ approach 1. Both cases reduce to a one-layer flow with a free surface, but the latter allows a velocity discontinuity in the interior of the fluid domain.

As a consequence of Eqs. (7), p and q are related by

$$q = \sqrt{H}\, p + F. \tag{9}$$

Combining these results, we conclude that the real characteristic speeds correspond to the solutions of the system

$$(p^2 - 1)(q^2 - 1) = \rho, \quad q = \sqrt{H}\, p + F, \tag{10}$$

with p, q all real. More precisely, each intersection point yielding a solution of this system corresponds to a real eigenvalue of A. Additionally, this geo-

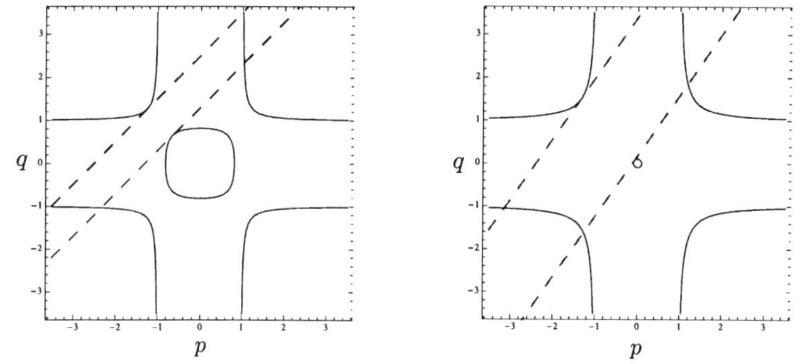

Fig. 3. Tangency condition for different physical parameters: $\rho = 1/3$, $H = 1$ (*left-hand side*) and $\rho = 99/100$, $H = 2$ (*right-hand side*). The solid and dashed lines represent (8) and (9) with $F = F^{\pm}_{crit}$, respectively.

metrical interpretation reveals that the system has at least two and a maximum of four real solutions. Hence, the system is of mixed type: it is strictly hyperbolic when we can present four real and distinct characteristics, and a system of composite type when there are two real and two imaginary characteristics, confirming the result obtained by using inner determinants. The passage between the two scenarios happens when the straight line described by Eq. (9) becomes tangent to the curve (8), as shown in Fig. 3. Notice that the intersection points with the boundary of the inner region representing the internal wave speeds disappear as the Froude number increases, leaving only the external (or surface) wave speeds, until it reaches F^{+}_{crit} beyond which the internal wave modes reappear.

For prescribed values of ρ and H, the curve (8) and the slope of the line (9) are completely determined. We seek the values of initial ordinates F for which the tangency holds. From Eq. (10), it follows that

$$H\,p^4 + 2\sqrt{H}F\,p^3 + (F^2 - H - 1)\,p^2 - 2\sqrt{H}F\,p - F^2 + (1 - \rho) = 0.$$

Multiple roots for this polynomial arise when its *discriminant* vanishes, leading to the condition $16H\,Q(y) = 0$. Surprisingly, this discriminant is precisely the same as Δ_7 in Eq. (5) and we now realize that the roots of $\Delta_7 = 0$ yield nothing but the condition for tangency. The results obtained can be summarized as follows:

Proposition 2.1. *For any physical parameters, there are two distinct positive real numbers* F^{-}_{crit} *and* F^{+}_{crit}, *with* $F^{-}_{crit} < F^{+}_{crit}$, *such that the system* (1) *is hyperbolic if and only if* $|F| \leqslant F^{-}_{crit}$ *or* $|F| \geqslant F^{+}_{crit}$.

3. Concluding remarks

Without solving the quartic equation, we have found explicit relations for which the system (1) is hyperbolic. In particular, in complete agreement with Ovsyannikov,[8] we perceive the effect caused by the presence of the free surface: the range of Froude numbers for hyperbolicity is not bounded, which is contrary to the rigid-lid case for which it can be shown[8] that *the characteristics for the rigid-lid system are real if and only if* $|F| \leqslant F_c$, where

$$F_c{}^2 = (1-\rho)(H+\rho)/\rho.$$

Numerical computations show that $F_c > F^-_{\text{crit}}$ and that these two values become almost indistinguishable in the Boussinesq limit. This could justify the common assumption that the difference between the rigid-lid and free-surface cases is insignificant, which we know is valid only for small relative speeds. Moreover, it is worth noticing that for small density ratios, F_c can actually exceed the value F^+_{crit}.

The well-posedness of the system was not addressed in this paper, but is a matter of great importance. It would be reasonable to expect that the system remains hyperbolic for initial data prescribed in the first range of Froude numbers. It is not clear yet if this is the case for the second branch.

The authors gratefully acknowledge support from NSF through Grant DMS-0620832 and ONR through Grant N00014-08-1-0377.

References

1. P. G. Baines, *Topographic Effects on Stratified Flows.* Cambridge Monographs on Mechanics (Cambridge University Press, 1995).
2. J. B. Schijf, J. C. Schnfeld, Theoretical considerations on the motion of salt and fresh water. In *Proc. of the Minn. Intl. Hyd. Conv. Joint meeting of the IAHR and Hyd. Div. ASCE*, pp. 321–333 (1953).
3. L. Armi, The hydraulics of two flowing layers with different densities. *J. Fluid Mech.*, **163**, pp. 27–58 (1986).
4. G. A. Lawrence, On the hydraulics of Boussinesq and non-Boussinesq two-layer flows. *J. Fluid Mech.*, **215**, pp. 457–480 (1990).
5. M. Castro, J. Macas, C. Pars, A Q-Scheme for a class of systems of coupled conservation laws with source term. Application to a two-layer 1-D shallow water system, *Math. Model. Numer. Anal.*, **35**, No. **1** pp. 107–127 (2001).
6. P. J. Montgomery, T. B. Moodie, Two-layer Gravity Currents with Topography. *Stud. Appl. Math.*, **102** pp. 221–266 (1999).
7. R. Barros, Conservation laws for one-dimensional shallow water models for one and two-layer flows. *Math. Models Meth. App. Sci.*, **16**, pp. 119–137 (2006).
8. L. V. Ovsyannikov, Two-layer "shallow water" model. *Journal of Applied Mechanics and Technical Physics*, **20**, Issue **2**, pp. 127–135 (1979).

9. V. V. Prasolov, *Polynomials* (New York: Springer, 2004).
10. E. I. Jury, M. Mansour, Positivity and nonnegativity conditions of a quartic equation and related problems. *IEEE Trans. Automat. Contr.*, **AC-26**, No. **2**, pp. 444–251 (1981).
11. A. T. Fuller, Root location criteria for quartic equations. *IEEE Trans. Automat. Contr.*, **AC-26**, No. **3**, pp. 777–782 (1981).

CONTRIBUTIONS TO BALANCED ARRAYS OF STRENGTH t WITH APPLICATIONS

D. V. CHOPRA

Department of Mathematics and Statistics, Wichita State Univ., Wichita, KS 67260-0033, USA
** E-mail: dharam.chopra@wichita.edu*

RICHARD M. LOW

Department of Mathematics, San Jose State University San Jose, CA 95192, USA
E-mail: low@math.sjsu.edu

Dedicated to Professor Daljit S. Ahluwalia on the occasion of his 75th birthday.

In this paper, we present some necessary existence conditions for balanced arrays (B-arrays) with two levels and having strength t (t being odd and even). We then specialize these conditions to some specific values of t to illustrate the usefulness and applications of these results.

Keywords: Balanced arrays; existence conditions.

1. Introduction and Preliminaries

For the sake of completeness, we first state some basic concepts and definitions. An *array* T with m *constraints* (corresponding to factors in design language), N *runs* (treatment-combinations), and with two *levels* is merely a matrix T of size $(m \times N)$ with two elements (say, 0 and 1). The symbols $P(\underline{\alpha})$, $\lambda(\underline{\alpha})$, and $w(\underline{\alpha})$ denote respectively (where $\underline{\alpha}$ is a column vector of a matrix T), the vector obtained by permuting the elements of $\underline{\alpha}$, the frequency with which $\underline{\alpha}$ occurs in T, and the *weight* of $\underline{\alpha}$ (the number of 1's in $\underline{\alpha}$). It is quite obvious that $w(\underline{\alpha}) = w[P(\underline{\alpha})]$. These arrays become very useful if we impose some combinatorial structure on them. One such combinatorial constraint leads to the following definition:

Definition 1.1. A matrix T of size $(m \times N)$ is called a *balanced* array

(B-array) of *strength* t ($t \leq m$) if for every ($t \times N$)-submatrix T^* of T, we have the following condition satisfied: $\lambda(\underline{\alpha}; T^*) = \lambda[P(\underline{\alpha}); T^*]$, where $\underline{\alpha}$ is a $(t \times 1)$ vector of T^*. If $w(\underline{\alpha}) = i$ $(0 \leq i \leq t)$, then the above condition can be written as $\lambda(\underline{\alpha}) = \lambda[P(\underline{\alpha})] = \mu_i$ (say). The vector $\underline{\mu}' = (\mu_0, \mu_1, \mu_2, \ldots, \mu_t)$ is called the *index set* of the array T, and we have $N = \sum_{i=0}^{t} \binom{t}{i} \mu_i$. Thus given $\underline{\mu}'$, we know N.

Definition 1.2. A B-array with $\mu_i = \mu$, for each i, is called an *orthogonal* array (O-array). In this case, we have $N = \mu \cdot 2^t$.

Thus, B-arrays include O-arrays as a special case. These arrays have been extensively used to construct fractional factorial designs in statistical design of experiments. It is well known that if $t = 2u + 2$ (t even), one can estimate all the interactions , up to and including $(u+1)$ factors, under the assumption that all interactions of more than $(u+1)$ factors are negligible. With an array of strength $t = 2u + 1$ (t odd), one is able to estimate all the interactions, up to and including u factors, even if interactions involving $(u + 1)$ factors are present. It was C.R. Rao [15] who first introduced O-arrays into statistics, under the name of hypercubes. It is not difficult to see that O-arrays with two elements only exist for $N = \mu \cdot 2^t$. Thus for $t = 4$, O-arrays may exist for $N = 16, 32, 48, \ldots$ In order to overcome this difficulty, the concept of B-arrays (under the name of partially balanced arrays) was introduced into statistics by I.M. Chakravarti [3] on the suggestion of C.R. Rao. An attractive feature of the designs constructed by using these arrays is that they are balanced. A design is said to be *balanced* if its variance-covariance matrix is invariant under a permutation of its factor symbols. Such designs lead to an ease in analysis and interpretation of results.

These combinatorial arrays have been primarily used in statistics in designing experiments, which is very important in almost all areas of scientific investigations such as industry, medicine, agriculture, etc. O-arrays, a special case of B-arrays, have been used in cryptography, computer science, error-correcting codes, information theory, and in the famous Taguchi techniques on quality control in industry. R.C. Bose [2] applied O-arrays to information theory to point out the connections between the problems of experimental designs and information theory. B-arrays are also related to (besides O-arrays) other combinatorial structures. For example, balanced incomplete block designs (BIBDS) are related to, in some fashion, B-arrays of strength two. Houghton, et. al [10] used this fact and computational techniques to show the non-existence of the famous BIB design $(46, 6, 1)$.

BIB designs have been quite useful in experimental situations where the number of treatments under study exceeds the block size. Sinha, *et al.* [21] have pointed out the relationship of B-arrays of strength two with rectangular designs, group divisible designs, and nested balanced incomplete block designs. Thus, B-arrays are not only useful in solving practical problems but also find great use theoretically in investigating other combinatorial structures. Hence, the existence and construction of such arrays are very important from the point of view of applications as well as to the study of other combinatorial structures. To gain further insight into the usefulness and importance of B-arrays to experimental designs and to combinatorics, the interested reader may consult the list of references (which, by no means, is an exhaustive one) at the end of the paper, and also further references mentioned therein.

The problem of constructing a B-array for a given set of parameters $\underline{\mu}'$ and m ($\geq t+1$) is clearly a non-trivial problem. To find the maximum number of constraints m for a given $\underline{\mu}'$ is an important problem from the point of view of combinatorics as well as that of design of experiments. Such problems for O-arrays have been discussed, among others, by Bose and Bush [1], C.R. Rao [15, 16], Seiden and Zemach [19], and for B-arrays by Chopra/Bsharat and or Dios [6, 7, 8], Rafter and Seiden [14], Saha *et al.* [18], Yamamoto *et al.* [21], etc.

In this paper, we derive some inequalities involving the parameters m and $\underline{\mu}'$ of a B-array with strength t. We consider cases when t is even (say, $t = 2u+2$), and when t is odd (say, $t = 2u+1$). For the existence of a B-array, it is necessary that each of these inequalities must be satisfied. We also describe the use of these inequalities to obtain $\max(m)$, for a given $\underline{\mu}'$. We illustrate the use of our results for some special values of t (t being odd or even).

2. Main Results with Discussion

The following results can be easily established.

Lemma 2.1. *A B-array with index set* $\underline{\mu}'$ *and* $m = t$ *always exists.*

Lemma 2.2. *A B-array* T *of strength* t *is also of lower strength* k $(\leq t)$.

Remark: It is obvious that the elements of the parameter vector of T, when considered as an array of strength k, are linear functions of the elements of $\underline{\mu}' = (\mu_0, \mu_1, \ldots, \mu_t)$. Let $A(j,k)$ be the jth $(0 \leq j \leq k)$ element of the parameter vector of T, when considered as an array of strength k,

$$A(j,k) = \sum_{i=0}^{t-k} \binom{t-k}{i} \mu_{i+j}, \quad \text{where } j = 0, 1, \ldots, k, \text{ and } k \leq t. \tag{1}$$

From (1), it follows that $A(t,t) = \mu_t$, $A(j,t) = \mu_j$, and $A(j,0) = N = A(0,0)$.

Lemma 2.3. *Consider a B-array T with m rows, strength t, and with index set $\underline{\mu}' = (\mu_0, \mu_1, \ldots, \mu_t)$. Let x_j $(0 \leq j \leq m)$ be the number of columns of weight j in T. Then, the following results hold:*

$$\sum_{j=0}^{m} \binom{j}{k} x_j = \binom{m}{k} A(k,k), \quad \text{where } k = 0, 1, 2, \ldots, t. \tag{2}$$

Remark: Considering T as an array of strength k, one can easily obtain Eq. (1) by counting in T the number of vectors of weight k through rows and columns.

Note: Let $S(a,b)$ denote the set of all the positive integers beginning with "a" and ending with "b", and let $S_k(a,b)$ denote the sum of the products of all the subsets of $S(a,b)$, each subset of k elements. Then, Eq. (2) can be rewritten as

$$\sum_{l=0}^{k} (-1)^{k-l} B_l S_{k-l}(1, k-1) = \sum_{l=0}^{k} (-1)^{k-l} m^l S_{k-l}(1, k-1) A(k,k), \tag{3}$$

where $B_l = \sum_{j=0}^{m} j^l x_j$, $l = 0, 1, \ldots, k$ and $k = 0, 1, \ldots, t$. For $l = k = 0$, the above is reduced to $B_0 = \sum_{j=0}^{m} x_j = A(0,0) = N$. We now make use of the above results to obtain existence conditions for B-arrays of strength t.

Theorem 2.1. *Consider a B-array T of strength $t = 2u + 1$ (say), with m rows and having index set $\underline{\mu}'$. Then for T to exist, the following condition must be satisfied:*

$$\sum_{k=0}^{2u} (-1)^k \binom{2u}{k} \frac{B_1^k}{N^k} B_{2u-k+1} \geq 0, \quad \text{where } B_k = \sum_{j=0}^{m} j^k x_j. \tag{4}$$

Proof. Clearly, $\sum_{j=0}^{m} j(j - \bar{j})^{2u} x_j \geq 0$, where $\bar{j} = \frac{\sum j x_j}{N} = \frac{B_1}{N}$. Thus, $\sum_{j=0}^{m} \sum_{k=0}^{2u} (-1)^k j \binom{2u}{k} j^{2u-k} (\bar{j})^k x_j \geq 0$. Substituting $\bar{j} = \frac{B_1}{N}$, we obtain $\sum_{j=0}^{m} \sum_{k=0}^{2u} (-1)^k \binom{2u}{k} \frac{B_1^k}{N^k} j^{2u-k+1} x_j \geq 0$. We can rewrite it as $\sum_{k=0}^{2u} (-1)^k \binom{2u}{k} \frac{B_1^k}{N^k} \sum_{j=0}^{m} j^{2u-k+1} x_j \geq 0$, which finally leads us to the inequality $\sum_{k=0}^{2u} (-1)^k \binom{2u}{k} \frac{B_1^k}{N^k} B_{2u-k+1} \geq 0$. This establishes the result. □

Theorem 2.2. *For a B-array T with m rows and of strength $t = 2u + 2$ $(u \geq 0)$ to exist, the following condition must be satisfied:*

$$\sum_{k=0}^{2u}(-1)^k \binom{2u}{k} B_{2u-k+2}\frac{B_1^k}{N^k} \geq 0. \tag{5}$$

Proof. In order to obtain Eq. (5), we consider $\sum j^2(j-\bar{j})^{2u}x_j \geq 0$, which leads to $\sum_{j=0}^{m}\sum_{k=0}^{2u} j^2(-1)^k\binom{2u}{k}j^{2u-k}(\bar{j})^k x_j \geq 0$ and the result follows. □

Theorem 2.3. *Let T be a B-array of strength $t = 2u$ with m rows, and index set $\underline{\mu}'$. Then for T to exist, the following condition must be satisfied:*

$$L_2L_{2u} \geq L_{u+1}^2 + L_2L_u^2, \quad \text{where} \tag{6}$$

$$L_2 = \sum_{k=0}^{1}(-1)^k N^{1-k}B_{2-k}B_1^k,$$

$$L_{2u} = \sum_{k=0}^{2u}(-1)^k\binom{2u}{k}N^{2u-k}B_{2u-k}B_1^k,$$

$$L_{u+1} = \sum_{k=0}^{u+1}(-1)^k\binom{u+1}{k}N^{u+1-k}B_{u+1-k}B_1^k \quad \text{and}$$

$$L_u = \sum_{k=0}^{u}(-1)^k\binom{u}{k}N^{u-k}B_{u-k}B_1^k,$$

where B_k is defined by $\sum j^k x_j$.

Proof. In order to establish the above result, we use from [13] the following result: $\alpha_{2u} \geq \alpha_{u+1}^2 + \alpha_u^2$ which refers to the moments of a set of data which has been standardized. For our data (which is the weight of the columns of T), the above inequality is $\sum_{j=0}^{m}(\frac{j-\bar{j}}{S})^{2u}x_j \geq [\sum(\frac{j-\bar{j}}{S})^{u+1}x_j]^2 + [\sum(\frac{j-\bar{j}}{S})^u x_j]^2$, where $\bar{j} = \frac{\sum jx_j}{N} = \frac{B_1}{N}$ and $S^2 = \frac{\sum(j-\bar{j})^2x_j}{N}$. Multiplying both sides by S^{2u+2}, we get $S^2\sum_{j=0}^{m}(j-\bar{j})^{2u}x_j \geq [\sum(j-\bar{j})^{u+1}x_j]^2 + S^2[\sum(j-\bar{j})^u x_j]^2$. Now, $S^2 = \frac{NB_2-B_1^2}{N^2}$. Using the binomial theorem to expand terms on both sides of the inequality, we obtain Eq. (6) after some simplification. □

Next, we take some special values of t to present some illustrative examples.

Example 1. For $t = 7$, we obtain from Eq. (4), $\sum_{k=0}^{6}(-1)^k\binom{6}{k}\frac{B_1^k}{N^k}B_{7-k} \geq 0$, which leads to $N^6B_7 - 6N^5B_6B_1 + 15N^4B_5B_1^2 - 20N^3B_4B_1^3 + 15N^2B_3B_1^4 - 6NB_2B_1^5 + B_1^7 \geq 0$. Let us take $\underline{\mu}' = (1,1,1,1,1,2,4,1)$, and check this inequality starting with $m = 8$. We find the inequality is contradicted when we use $m = 10$. Hence, max(m) for this array is ≤ 9. However, using results given in [7], we obtain max$(m) \leq 11$.

Example 2. For $t = 6$, we use Eq. (5) by substituting $u = 2$ in it to obtain $N^4B_6 - 4N^3B_5B_1 + 6N^2B_4B_1^2 - 4NB_3B_1^3 + B_2B_1^4 \geq 0$. This gives us max$(m) \leq 15$, for $\underline{\mu}' = (1,1,3,1,1,2,2)$. Using Eq. (6) with $u = 3$, we find max(m) is much greater than 15.

Thus, the results presented here permit us to study the existence of a B-array for a given m, $\underline{\mu}'$, and any t. Also, these inequalities allow us to find the max(m) for any strength t array, for a given $\underline{\mu}'$. Our investigations also lead us to believe that there is no one single existence condition which is uniformly better than others, for all B-arrays.

References

1. R.C. Bose and K.A. Bush, Orthogonal arrays of strength two and three, *Ann. Math. Statist.* **23** (1952), 508–524.
2. R.C. Bose, On some connections between the design of experiments and information theory, *Bull. Internat. Statist. Inst.* **38** (1961), 257–271.
3. I.M. Chakravarti, Fractional replication in asymmetrical factorial designs and partially balanced arrays, *Sankhya* **17** (1956), 143–164.
4. C.S. Cheng, Optimality of some weighing and 2^m fractional designs, *Ann. Statist.* **8** (1980), 436–444.
5. D.V. Chopra, Balanced optimal 2^8 fractional factorial designs of resolution V, $52 \leq N \leq 59$, *A Survey of Statistical Designs and Linear Models*, North-Holland Publishing Co., Amsterdam (1975), 91–99.
6. D.V. Chopra, On balanced arrays with two symbols, *Ars Combin.* **20A** (1985), 59–63.
7. D.V. Chopra and M. Bsharat, On the existence of balanced arrays of strength seven, *Congr. Numer.* **173** (2005), 169–174.
8. R. Dios and D.V. Chopra, Investigations on the existence of some balanced arrays with two symbols, *J. Combin. Math. Combin. Comput.* **58** (2006), 33–39.
9. A.S. Hedayat, N.J.A. Sloane and J. Stufken, *Orthogonal Arrays: Theory and Applications* (1999), Springer-Verlag (N.Y.).
10. S.K. Houghton, I. Thiel, J. Jansen and C.W. Lam, There is no (46, 6, 1) block design, *J. Combin. Designs* **9** (2001), 60-71.
11. J.P.C. Kleijnen and Ozge Pala, Maximizing the simulation output: a compe-

tition, *Simulation* **7** (1999), 168–173.

12. J.Q. Longyear, Arrays of strength t on two symbols, *J. Statist. Plann. Inf.* **10** (1984), 227–239.
13. D.S. Mitrinovic, *Analytic Inequalities* (1970), Springer-Verlag (N.Y.).
14. J.A. Rafter and E. Seidon, Contributions to the theory and construction of balanced arrays, *Ann. Statist.* **2** (1974), 1256–1273.
15. C.R. Rao, Hypercubes of strength d leading to confounded designs in factorial experiments, *Bull. Calcutta Math. Soc.* **38** (1946), 67–78.
16. C.R. Rao, Factorial experiments derivable from combinatorial arrangements of arrays, *J. Roy. Statist. Soc. Suppl.* **9** (1947), 128–139.
17. C.R. Rao, Some combinatorial problems of arrays and applications to design of experiments, *A Survey of Combinatorial Theory* (edited by J.N. Srivastava, et. al.) (1973), North-Holland Publishing Co., 349–359.
18. G.M. Saha, R. Mukerjee and S. Kageyama, Bounds on the number of constraints for balanced arrays of strength t, *J. Statist. Plann. Inf.* **18** (1988), 255–265.
19. E. Seiden and R. Zemach, On orthogonal arrrays, *Ann. Math. Statist.* **27** (1966), 1355–1370.
20. K. Sinha, V. Dhar, G.M. Saha and S. Kageyama, Balanced arrays of strength two from block designs, *Combin. Designs* **10** (2002), 303–312.
21. S. Yamamoto, M. Kuwada and R. Yuan, On the maximum number of constraints for s-symbol balanced arrays of strength t, *Commun. Statist. Theory Meth.* **14** (1985), 2447-2456.

ASYMPTOTIC SOLUTIONS OF SOME RANDOMLY PERTURBED NONLINEAR WAVE EQUATIONS

PAO-LIU CHOW

Department of Mathematics, Wayne State University, Detroit, Michigan 48202, USA
E-mail: plchow@math.wayne.edu

The paper is concerned with the long-time behavior of solutions to some nonlinear stochastic wave equations. In particular we are interested the questions of local and global solutions and the asymptotic solutions as $t \to \infty$. Under appropriate conditions, the theorems on the existence of local and global solutions as well as the invariant distributions will be presented.

Keywords: Stochastic Wave equation, Polynomial nonlinearity, Asymptotic Solution, Invariant Distribution.

1. Introduction

Consider the nonlinear Klein-Gordon equation perturbed by a state-dependent white noise:

$$\begin{cases} \partial_t^2 u(x,t) = c^2 \nabla^2 u - 2\alpha \partial_t u - \gamma u + f(u) + \sigma(u)\dot{W}(x,t), & t > 0, \\ u(x,0) = g(x), \quad \partial_t u(x,0) = h(x), & x \in \mathbb{R}^d, \end{cases} \tag{1}$$

where c, α, γ are the wave speed, the damping coefficient and the dispersion parameter, respectively; $f(u)$ and $\sigma(u,x,t)$ are some nonlinear functions, and $g(x), h(x)$ are given initial data. $\dot{W}(x,t) = \partial_t W(x,t)$ is spatially dependent White Noise, and $W(x,t)$ is a Gaussian (Wiener) random field such that $\mathbb{E}\{W(x,t)\} = 0, \quad \mathbb{E}\{W(x,t)W(y,s)\} = \min\{t,s\} r(x,y)$, where $\mathbb{E}$ denotes the mathematical expectation and $r(x,y)$ is the spatial correlation function. In the deterministic case ($f \equiv 0$) without the damping ($\alpha = 0$), the equation (1) reduces to

$$\begin{cases} \partial_t^2 u(x,t) = c^2\nabla^2 u - \gamma u + f(u), & t > 0, \\ u(x,0) = g(x), \quad \partial_t u(x,0) = h(x), & x \in \mathbb{R}^d. \end{cases} \tag{2}$$

It was first shown by J.B.Keller[4] in 1957 that, for a certain class of $f(u)$ with $d \leq 3$, the solution $u(x,t)$ becomes unbounded in a finite time t_e so that $\lim_{t\to t_e^-} \sup_x |u(x,t)| = +\infty$, provided that the initial data satisfy appropriate conditions. It is conceivable that the blowup of solutions in a finite time may occur in the stochastic wave equation like Eq. (1) as well. This has lead us to study the questions concerning the existence of a global solutions and the asymptotic behavior of solutions as $t \to \infty$.

As to be seen, the main tools of analysis are based on the energy method and the Itô calculus for stochastic partial differential equations.[3] The paper is organized as follows. In section 2, the energy equation and an energy inequality will be given. The existence of an explosive solution for a nonlinear stochastic wave equation will be demonstrated by an example in section 3. Then, in section 4, an existence theorem for a global solution to a stochastic wave equation with a polynomial nonlinearity will be presented. Finally, in Section 5, we shall discuss the problem of asymptotic solutions as $t \to \infty$ and the associated equilibrium distributions.

2. Energy Equation and Inequality

Consider the linear equation in the domain $\mathcal{D} \subset \mathbb{R}^d$ with a smooth boundary $\partial\mathcal{D}$:

$$\begin{cases} \partial_t^2 u(x,t) = c^2\nabla^2 u - 2\alpha\partial_t u - \gamma u + f(x,t) + \sigma(x,t)\dot{W}(x,t), \\ u(x,0) = g(x), \quad \partial_t u(x,0) = h(x), \quad x \in \mathbb{R}^d \\ \quad u|_{\partial\mathcal{D}} = 0, \end{cases} \tag{3}$$

where $f(x,t)$ and $\sigma(x,t)$ are non-anticipating random fields such that $\mathbb{E}\int_0^t \{\|f_s\|^2 + Q_s\}\,ds < \infty$, with $Q_s = \int r(x,x)\sigma^2(\cdot,s)\,dx$. By applying the Itô formula[3] to $\|u_t\|^2$ with the aid of Eq. (3), one can derive the energy equation:

$$
\begin{aligned}
\mathbf{e}(u_t) = \mathbf{e}(u_0) - \alpha \int_0^t \|v_s\|^2 ds + \int_0^t (v_s, f_s)\, ds \\
+\frac{1}{2}\int_0^t Q_s\, ds + \int_0^t (v_s, \sigma_s\, dW_s), \quad \text{a.s.}
\end{aligned}
\tag{4}
$$

where the energy (density) function is defined as

$$
\mathbf{e}(u_t) = \{\|v_t\|^2 + \gamma\|u_t\| + c^2\,\|Du_t\|\,\}. \tag{5}
$$

Introduce the energy norm $\|u_t\|_e = \{\mathbf{e}(u_t)\}^{1/2}$, which is equivalent to the norm of $(u_t; v_t)$ in $H_0^1 \times H$: $\|(u_t; v_t)\| = \{\|v_t\|^2 + \|u_t\|_1^2\}^{1/2}$, where $\|u_t\|_1^2 = \{\|u_t\|^2 + \|Du_t\|^2\}$ is the Sobolev norm of first order.

To obtain an exponential estimate, we need to introduce the pseudo energy function

$$
\mathbf{e}^\lambda(u_t) = \frac{1}{2}\{\|v_t + \lambda u_t\|^2 + \|u_t\|^2 + \|Du_t\|^2\}, \quad \lambda > 0. \tag{6}
$$

It is not hard to show these two energy functions are equivalent in the sense that $C_1\,\mathbf{e}(u_t) \le \mathbf{e}^\lambda(u_t) \le C_2\,\mathbf{e}(u_t)$, for some $C_2 > C_1 > 0$. By combining PDE techniques and the Itô calculus, one can obtain the following key lemma (see Lemma 2.2 in the paper[2]).

Lemma 2.1. *Let u_t be the strong solution of the linear problem (3). Then there exist positive constants λ_0, η with $0 < \eta < \lambda < \lambda_0$ such that the following energy inequality holds*

$$
\mathbb{E}\mathbf{e}^\lambda(u_t) \le \mathbf{e}^\lambda(u_0)e^{-\eta t} + \int_0^t e^{-\eta(t-s)}\mathbb{E}\{\ \frac{2}{\eta}\|f_s\|^2 + Q_s\ \}ds. \tag{7}
$$

3. Example of Unbounded Solution

To demonstrate the existence of an explosive solution, we consider the following cubically nonlinear stochastic wave in $\mathbb{R}^3$:

$$
\begin{cases}
\partial_t^2 u = c^2\nabla^2 u - \gamma\, u + k\, u^3 + \partial_t M(u,x,t), \quad t > 0 \\
u(x,0) = g(x), \quad \partial_t u(0,x) = h(x), \quad x \in \mathbb{R}^d, \ d \le 3, \\
M(u,x,t) = \int_0^t \sigma(u(x,s),x,s)dW(x,s), \ k > 0.
\end{cases}
\tag{8}
$$

The corresponding system of Itô equations is given by

$$\begin{cases} du_t = v_t dt, \\ dv_t = c^2\nabla^2 u_t\, dt - \gamma\, u_t\, dt + k\, u_t^3\, dt + d\, M_t(u), \\ u_0 = g, \quad v_0 = h. \end{cases} \tag{9}$$

We will show that the mean-square of the solution of Eq. (8) may become unbounded in a finite time in the L^2-sense. To this end we let

$$\varphi(t) = \frac{1}{2}\mathbb{E}\|u_t\|^2, \quad F_\alpha(t) = \varphi^{-\alpha}(t), \text{ for } \alpha > 0. \tag{10}$$

Then we can compute the first two derivatives of $\phi(t)$ as follows:

$$\begin{aligned} \varphi^{'}(t) &= (g, h) + \mathbb{E}\int_0^t \{\|v_s\|^2 - \gamma\,\|u_s\|^2 - c^2\|Du_s\|^2 + k\,\|u_s^2\|^2\}ds, \\ \varphi^{''}(t) &= \mathbb{E}\left\{\|v_t\|^2 - \gamma\,\|u_t\|^2 - c^2\|Du_t\|^2 + k\,\|u_t^2\|^2\right\}. \end{aligned} \tag{11}$$

Instead of showing $\Phi(t) \to \infty$ as t tends to some $t \to t_e > 0$, we will show that $F_\alpha(t) \to 0$ as $t \to t_e$. The idea of proof is to show that the graph of the function $F_\alpha(t)$ intersects the t-axis by means of its first two derivatives. By computing the derivatives and making use of the Eq. (11), we can get

$$\begin{aligned} F_\alpha^{'}(t) &= -\alpha\varphi^{-(\alpha+1)}(t)\varphi^{'}(t), \\ F_\alpha^{''}(t) &= \alpha\varphi^{-(\alpha+1)}(t)\left\{(\alpha+1)\frac{[\varphi^{'}(t)]^2}{\varphi(t)} - \varphi^{''}(t)\right\} \\ &\leq \alpha\varphi^{-(\alpha+1)}(t)G_\alpha(t), \end{aligned}$$

where

$$\begin{aligned} G_\alpha(t) &= (2\alpha+1)\mathbb{E}\|v_t\|^2 + \mathbb{E}\{\gamma\|u_t\|^2 + c^2\,\|Du_t\|^2 - k\,\|u_t^2\|^2\} \\ &= -(2\alpha+1)\{\tfrac{k}{2}\|u_0^2\|^2 - e(u_0) - \textstyle\int_0^t Q_s ds\} \\ &\quad -2\alpha\mathbb{E}(c^2\|Du_t\|^2 + \gamma\,\|u_t\|^2) - k\,(\frac{1}{2} - \alpha)\mathbb{E}\|u_t^2\|^2. \end{aligned}$$

Choose $\alpha \in (0, 1/2)$ so that the above yields

$$G_\alpha(t) < -(2\alpha+1)\left\{\frac{k}{2}\|u_0^2\|^2 - \mathbf{e}(u_0) - \int_0^t Q_s ds\right\}.$$

Suppose that

(i) $u_0 \in C_b^1 \cap L^4$ and $v_0 \in H$ such that $(u_0, v_0) > 0$.

(ii) $q_0 = \int_0^\infty \int r(x,x)\sigma_0^2(x,t)dx\,dt < \infty$, where $\sigma_0^2(x,t) = \sup_u\{\sigma^2(u,x,t)\}$.

(iii) $k > 2[q_0 + \mathbf{e}(u_0)]/\|u_0^2\|^2$.

Then we have $G_\alpha(t) < 0$ for any $t \geq 0$, and moreover,

(1) $F_\alpha(0) = F_\alpha(0) = \varphi^{-\alpha}(0) = (\frac{1}{2}\|u_0\|^2)^{-\alpha} > 0$.

(2) $F_\alpha'(0) = -\alpha\varphi^{-(\alpha+1)}(0)\varphi'(0)$
$= -\alpha(\frac{1}{2}\|u_0\|^2)^{-(\alpha+1)}(u_0, v_0) < 0$.

(3) $F_\alpha''(t) \leq \alpha\varphi^{-(\alpha+1)}(t)G_\alpha(t) < 0$ for $t \geq 0$.

It follows from the results (1)–(3) given above that there is $t_e < T_0 = F_\alpha(0)/|F_\alpha'(0)|$ such that $\lim_{t\to t_e^-} F_\alpha(t) = \lim_{t\to t_e^-} \varphi^{-\alpha}(t) = 0$ or $\lim_{t\to t_e^-} \varphi(t) = \frac{1}{2}\lim_{t\to t_e^-} \mathbb{E}\,\|u_t\|^2 = \infty$. This shows that, under conditions (i)-(iii), the solution blows up at t_e as asserted. We remark that it is possible to generalize the above result to a more general class of nonlinear function $f(u)$.

3.1. *Existence of Global Solutions*

Consider the following Cauchy problem in a domain $\mathcal{D}$:

$$\begin{cases} \partial_t^2 u = c^2\nabla^2 u - 2\alpha\partial_t u - \gamma u + f(u,x) + \sigma(Ju,x,t)\partial_t W(x,t) \\ u(x,0) = g(x),\ \partial_t u(x,0) = h(x),\ x \in \mathbb{D}, \\ u|_{\partial\mathbb{D}} = 0, \end{cases} \tag{12}$$

where $\mathcal{D} \subset \mathbb{R}^d$ with $d \leq 3$, $f(u,\cdot)$ and $\sigma(Ju,\cdot,t)$ are some nonlinear functions of u and $Ju = (u, \partial_{x_1}u, \ldots, \partial_{x_d}u)$. The corresponding Itô equations read:

$$\begin{cases} du_t = v_t dt \\ dv_t = (c^2\nabla^2 - 2\alpha v - \gamma)u_t dt - 2\alpha v_t dt + f_t(u_t)dt + \sigma_t(Ju_t)dW_t \\ u_0 = g, \ v_0 = h. \end{cases} \tag{13}$$

For the existence of global solutions, we impose the following conditions:

(A1) $f(u,x) = \sum_{j=1}^{m} a_j(x)u^j$, and
$G(u,x) = -2\int_0^u f(s,x)dx \geq (b_1 + b_2 u^{2k})u^2$, for $x \in \mathbb{R}^d$, $u \in \mathbb{R}$ and $b_1 \geq 0$, $b_2 > 0$,
where $m = 2k+1$ with $k = 1$ for $d = 3$ and $k \geq 0$ for $d = 1$ or 2, and $a_j(x)$ is bounded and continuous for $j = 1, ..., m$.

(A2) The function $\sigma(s,y,x,t)$ is continuous such that

$$|\sigma(s,y,x,t)|^2 \leq C_1(1 + |s|^{2(k+1)} + |y|^2), \ \forall t \geq 0, \ x, y \in \mathbb{R}^d,$$

for some constant $C_1 > 0$.

(A3) $$\left|\sigma(s,y,x,t) - \sigma(s',y',x,t)\right|^2 \leq C_2\left\{(1 + |s|^{2k} + \left|s'\right|^{2k})\left|s - s'\right|^2 + \left|y - y'\right|^2\right\}$$
for some $C_2 > 0$, $\forall\, t \geq 0$, $x \in \mathbb{R}^d$; $s, s' \in \mathbb{R}$; $y, y' \in \mathbb{R}^d$.

(A4) $W(t,x)$, $t \geq 0, x \in \mathbb{R}^d$, is a continuous Wiener random field with covariance function $r(x,y)$ such that

$$Tr\ R = \int r(x,x)dx < \infty, \quad \text{with } r_0 = \sup_{x \in \mathbb{R}^d} r(x,x) < \infty.$$

Then it was proved in the paper[1] that the following theorem holds true.

Theorem 3.1. *If the conditions (A1)–(A4) hold, then, for $u_0 \in H_1, v_0 \in H$, the Cauchy problem (12) has a unique continuous global solution $u(\cdot,t) \in H_1$ with $\partial_t u(\cdot,t) \in H$. Moreover the solution is bounded uniformly in mean-square so that $\sup_{t\geq 0} \mathbb{E}\{\mathbf{e}(u_t)\} < \infty$, where $\mathbf{e}(u_t) = \frac{1}{2}\{\|\partial_t u_t\|^2 + c^2\|Du_t\|^2 + \gamma\|u_t\|^2\}$.*

The proof is based on a H^1-Lipschitz truncation technique to obtain a local solution for the regularized equation. Then, by condition (A4) and the energy inequality in Lemma 2.1, we can remove the cut-off and extend the time interval for the local solution to $[0, \infty)$.

4. Asymptotic Solutions

Before studying the long-time asymptotic solution, let us consider the following simple example:

$$\begin{cases} \partial_t^2 u = c^2 \partial_x^2 u - 2\alpha \partial_t u_t + \partial_t W(x,t), \;\; t > 0, \;\; x \in \mathcal{D} = (0,\pi), \\ u(x,0) = h(x), \;\; \partial_t(x,0) = 0, \\ u(0,t) = u(\pi,t) = 0, \end{cases} \tag{14}$$

where $W(x,t) = \sum_{n=1}^{\infty} \sigma_n b_n(t)\varphi_n(x)$ with $\sum_{n=1}^{\infty} \sigma_n^2 < \infty$, $\{b_n(t)\}$ is sequence of independent of standard Brownian motions, and $\varphi_n = \sqrt{2/\pi}\sin nx$, $n = 1, 2, \cdots$, are the normalized eigenfunctions associated with this problem. Then, by the method of eigenfunctions expansion,[3] the problem (14) can be easily solved to give

$$u(x,t) = \sum_{n=1}^{\infty} \{h_n \cos\omega_n t e^{-\alpha t} + \frac{\sigma_n}{\omega_n} \int_0^t e^{-\alpha(t-s)} \sin\omega_n(t-s) db_n(s)\}, \tag{15}$$

where $u_n(t)\varphi_n(x)$, $h(x) = \sum_{n=1}^{\infty} h_n \varphi_n(x)$, and $\omega_n = \sqrt{(nc)^2 - \alpha^2}$ for $n = 1, 2, \cdots$.. In view of Eq. (15), the solution u_t is a Gaussian random field. As $t \to \infty$, $u(x,t)$ converges in mean-square to a Gaussian random function $\varphi(x)$ with mean zero and the covariance function q(x,y) so that $\mathbb{E}\{u(x,t)u(y,t)\} \sim q(x,y) = \sum_{n=1}^{\infty} \frac{1}{4\alpha}(\frac{\sigma_n}{nc})^2 \varphi_n(x)\varphi_n(y)$. The probability distribution μ of φ is the invariant measure for this problem.

Consider the autonomous version of Eq. (13):

$$\begin{cases} du_t = v_t dt, \\ dv_t = (c^2\nabla^2 u_t - \gamma u_t - 2\alpha v_t)\, dt + F(u_t)dt + \Sigma(u_t)dW_t \\ u_0 = g, \; v_0 = h, \end{cases} \tag{16}$$

where $F(u) = f(Ju, \cdot)$, $\Sigma(u) = \sigma(Ju, \cdot)$ and $Ju = (u; Du)$. Assume that

(B1) There exis constants b_i, c_i for $i = 1, 2$, such that $\|F(u)\|^2 \leq b_1\|u\|_1^2 + c_1$, $\|\Sigma(u)\|_R^2 \leq b_2\|u\|_1^2 + c_2$, for any $u \in H_1$, where $\|\sigma\|_R^2 = \int r(x,x)\,\sigma^2(x)\,dx$.

(B2) There exist constants $k_1, k_2 > 0$ such that

$$\|F(u) - F(u')\|^2 \leq k_1 \|u - u'\|_1^2, \ \|\Sigma(u) - \Sigma(u')\|_R^2 \leq k_2 \|u - u'\|_1^2,$$

for any $u, u' \in H_1$.

(B3) The constants b_i and k_i satisfy $(b_1 + b_2\lambda) \wedge (k_1 + k_2\lambda) \leq \frac{\lambda^2}{2}$, for $\lambda < \lambda_0$.

Theorem 4.1. *Suppose that the conditions (B1)–(B3) are satisfied. Then, for any $u_0 \in H_1, v_0 \in H$, the solution u_t of the Cauchy problem (14) converges, as $t \to \infty$, to a random function φ in mean-square with $\mathbb{E}\,\|\varphi\|^2 < \infty$. The probability distribution $\mathcal{L}\{\varphi\}$ is the unique invariant measure μ for this problem.*

The main idea of proof is to extend Eq. (16) to the following

$$\begin{cases} d\widetilde{u}_t = \widetilde{v}_t dt, \\ d\widetilde{v}_t = (c^2\nabla^2\widetilde{u}_t - \gamma\widetilde{u}_t - 2\alpha\widetilde{v}_t)\,dt + F(\widetilde{u}_t)dt + \Sigma(\widetilde{u}_t)d\widetilde{W}_t \\ \widetilde{u}_\tau = g, \ \widetilde{v}_\tau = h, \quad t > \tau > -\infty, \end{cases} \tag{17}$$

where $\widetilde{W}_t = W_t$ for $t \geq 0$ and $\widetilde{W}_t = V_{-t}$ for $t \leq 0$], and W_t and V_t be two independent, identically distributed Wiener random fields. With the aid of the energy inequality (Lemma 2.1), it can be shown that, for any $t > 0$, the solution $u_t(\tau)$ of Eq. (17) converges in $L^2(\Omega, H)$ to a random function φ as $\tau \to -\infty$. This implies that the solution of Eq. (16) $u_t = u_0(-t)$ (in distribution) converges to φ. The probability distribution of φ is the invariant measure μ mentioned in the theorem. The detailed proof can be found in the paper.[3]

References

1. P.L. Chow, *Stochastic wave equations with polynomial nonlinearity*, Annals Appl. Probab. **12** (2002), 361-381.
2. P.L. Chow, *Asymptotics of solutions to semilinear stochastic wave equations*, Annals Appl. Probab. **16** (2006), 757-789.
3. P.L. Chow, *Stochastic Partial Differential Equations*, Chapman and Hall/CRC, Boca Raton, London, New York, 2007.
4. J.B. Keller, *On solutions of nonlinear wave equations* Comm. Pure and Appl. Math., **10** (1957), 523–530.

THE BOOTSTRAP IN BINARY MODEL DIAGNOSTICS

GERHARD DIKTA

Department of Medizintechnik und Technomathematik,
Aachen University of Applied Sciences,
Jülich, D-52428, Germany
E-mail: dikta@fh-aachen.de
www.fh-aachen.de/dikta_g.html

Consider a binary regression model, where the conditional expectation of the binary variable given an explanatory variable belongs to a parametric family. To check whether a sequence of independent and identically distributed observations of these variables belongs to such a parametric family, we use Kolmogorov-Smirnov and Cramér-von Mises type tests which are based on maximum likelihood estimation of the parameter and on a marked empirical process introduced by Stute. We study a new resampling scheme of the bootstrap in this setup to approximate the critical values belonging to these tests. Furthermore, this approach is applied to simulated and real data. In the latter case we check parametric model assumptions of some right censored data sets. These checks are necessary if one wants to apply a semiparametric approach to analyze the censored data.

Keywords: Bootstrap, binary models, marked empirical process, model diagnostic

1. Introduction

Consider a sequence of independent and identically distributed (i.i.d.) data

$$(\delta_1, Z_1), \ldots, (\delta_n, Z_n),$$

where δ is Bernoulli distributed and Z is distributed on $\mathbb{R}$ or $\mathbb{R}_{\geq 0}$ with a continuous distribution function (d.f.) H. For notational reasons we assume here that Z is concentrated on $\mathbb{R}_{\geq 0}$.

This type of observations occurs in different fields of application. In a clinical study, for example, Z can be a baseline value of a patient at screening time and δ might be an indicator showing the success of a certain therapy at the end of the study. As a second example, consider data from a survival study, where the observations are generated under the random censorship model (RCM). Under RCM one has two independent i.i.d. sequences: the survival times $X_1, \dots, X_n$ and the censoring times $Y_1, \dots, Y_n$. One observes $Z_i = \min(X_i, Y_i)$ and an indicator $\delta_i = I(X_i \leq Y_i)$ showing whether the corresponding Z_i–observation is censored or not.

To analyze such data statistically, one often assumes (in accordance with an expert) a parametric regression model for

$$m(z) = \mathbb{E}\big(\delta \,|\, Z = z\big),$$

the conditional expectation of δ given $Z = z$. Precisely, one assumes that m belongs to a known parametric family $\mathcal{M} := \{m(\cdot, \theta) \,|\, \theta \in \Theta\}$, where $\Theta \subset \mathbb{R}^k$ denotes the parameter space, i.e. $m(\cdot) = m(\cdot, \theta_0)$, for some $\theta_0 \in \Theta$.

Obviously, any conclusion based on such a parametric model depends on the validity of the model. Therefore, it is extremely important to check the data for a departure from the model. Precisely, one has to check the null hypothesis

$$H_0 : m \in \mathcal{M} \;\text{ versus }\; H_1 : m \notin \mathcal{M}.$$

Rather than concentrating on an optimal procedure for a specific parametric model, our objective is to setup an universal approach, which might not result in an optimal test for a specific parametric model but which is capable to handle a wide range of parametric models.

To fit a parametric regression function, maximum likelihood estimation (MLE) is usually applied to estimate the parameter θ_0 by θ_n, that is,

$$\theta_n = \arg\max_{\theta \in \Theta} l_n(\theta),$$

where

$$l_n(\theta) = \frac{1}{n} \sum_{i=1}^{n} \Big(\delta_i \log\big(m(Z_i, \theta)\big) + (1 - \delta_i) \log\big(1 - m(Z_i, \theta)\big)\Big)$$

is the normalized log-likelihood function.

Our method is based on Stute's[1] functional central limit theorem (FCLT) for a marked empirical process (MEP). In the case of binary data, the MEP is defined by

$$R_n(z) = n^{-1/2} \sum_{i=1}^{n} (\delta_i - m(Z_i, \theta_n)) \cdot I(Z_i \leq z), \quad 0 \leq z \leq \infty,$$

where I denotes the indicator function. According to Stute's[1] FCLT,

$$R_n \longrightarrow R_\infty \quad \text{in distribution in } D([0, \infty]),$$

if H_0 is correct. Here, $D([0,\infty])$ denotes the Skorokhod space. R_∞ is a centered Gaussian process with a complicate covariance structure depending on $\mathcal{M}$, on θ_0, and on H. Based on this result, an application of the continuous mapping theorem guaranties the convergence in distribution of Kolmogorov-Smirnov (KS)

$$D_n = \sup_{0\le z\le\infty} |R_n(z)| \longrightarrow \sup_{0\le z\le\infty} |R_\infty(z)|$$

and Cramér-von Mises (CvM)

$$W_n = \int R_n^2\, dH_n \longrightarrow \int R_\infty^2\, dH$$

type test statistics. H_n denotes the empirical distribution function (edf) of the $Z-$sample. However, the limit distributions of these two tests depend on the covariance structure, on the model, and on the true parameter. Therefore, critical values and p-values are not tractable in general anymore. But this is a situation, where the bootstrap might be a promising approach.

2. The Bootstrap Approach

Assume that the bootstrap sample is given by an i.i.d. sequence

$$(\delta_1^*, Z_1^*), \ldots, (\delta_n^*, Z_n^*)$$

generated under some resampling scheme, which will be specified later. Based on $\theta_n^* = \arg\max_{\theta\in\Theta} l_n^*(\theta)$, the corresponding MLE belonging to the bootstrap data, the bootstrap version of the MEP is then defined by

$$R_n^*(z) = n^{-1/2}\sum_{i=1}^n (\delta_i^* - m(Z_i^*, \theta_n^*)) \cdot I_{\{Z_i^*\le z\}}, \quad 0\le z\le\infty.$$

If for $\mathbb{P}-$a. e. sample sequence

$$R_n^* \longrightarrow R_\infty \quad \text{in distribution in } D([0,\infty])$$

holds, the critical values of the KS and of the CvM test can be approximated through

$$D_n^* = \sup_{0\le z\le\infty} |R_n^*(z)| \quad \text{and} \quad W_n^* = \int (R_n^*)^2\, dH_n^*,$$

respectively. Here H_n^* denotes the edf of the Z^*-sample.

To apply a bootstrap approach in this setup is not new. Stute *et al.*[2] used it to check for a regression model in a different framework. To check for a parametric binary model, Zhu *et al.*[3] used a classical bootstrap approach

(CB), that is, each pair (δ^*, Z^*) of the bootstrap sample is uniformly distributed on the original data set (ODS). They proved that under the CB resampling scheme

$$\bar{R}_n^* = R_n^* - R_n \longrightarrow R_\infty \quad \text{in distribution in } D([0,T]),$$

where $H(T) < 1$. If the null hypothesis is correct, this convergence holds for $\mathbb{P}-$a.e. sample sequence. Nevertheless, a correction term R_n has to be substracted from R_n^*, to obtain the desired limiting process. Furthermore, the Skorokhod space is restricted to a compact interval $[0, T]$ with positive mass to the right of T.

To approximate the distribution of the original statistic in a hypothesis test by bootstrap, one has to guarantee that the critical values or p-values obtained from the bootstrap are meaningful, regardless whether the original data are generated under H_0 or H_1 (see Efron and Tibshirani,[4] p. 232). In the case of the CB resampling scheme, the bootstrap does not satisfy this requirement.

To face this necessity, we use the following model-based (MB) resampling scheme

Definition 2.1 (MB resampling scheme). *Based on the original i.i.d. data* $(\delta_1, Z_1), \ldots, (\delta_n, Z_n)$ *and the corresponding MLE* θ_n*, the model-based resampling scheme is defined by:*

(a) Set $Z_i^* = Z_i$*, for* $1 \le i \le n$.

(b) Generate a sample $\delta_1^*, \ldots, \delta_n^*$ *of independent Bernoulli random variables, where* δ_i^* *has probability of success given by* $m(Z_i, \theta_n)$.

Under the MB resampling scheme Dikta *et al.*[5] obtained the following

Theorem 2.1 (Main result). *Under some general assumptions, if the null hypothesis is correct then, for* $\mathbb{P}-$*a.e. sample sequence,*

$$R_n^* \longrightarrow R_\infty \quad \text{in distribution in } D([0,\infty]), \tag{1}$$

if MB is the underlying resampling scheme. Furthermore, if the alternative is correct, Eq. (1) still holds under some general assumptions and an adapted interpretation of the parameter θ_0.

For the proof and the concrete assumptions of Theorem 2.1 the reader is referred to Dikta *et al.*[5]

Based on Theorem 2.1, approximated p-values of KS and CvM test statistics can now be obtained. The following algorithm shows the steps to calculate an approximated p-value for the KS statistic D_n.

Algorithm 2.1 (Approximation of a p-value). *Calculate the MLE θ_n and D_n corresponding to the ODS $(\delta_1, Z_1), \ldots, (\delta_n, Z_n)$.*

- *Use the MB resampling scheme to generate N independent i.i.d. bootstrap samples (N replications)*

$$(\delta^*_{1,k}, Z^*_{1,k}) \ldots, (\delta^*_{1,k}, Z^*_{1,k}), \quad 1 \leq k \leq N$$

and calculate for each bootstrap sample $D^*_{n,k}$.

- *Obtain an approximated p-value for D_n by*

$$\frac{1}{N} \sum_{k=1}^{N} I(D^*_{n,k} > D_n).$$

Similar algorithms can be used for the CvM statistic under the MB resampling scheme. In the case of the CB resampling scheme, one has to use $\bar{R}^*_n$ instead of R^*_n to obtain $D^*_{n,k}$ and $W^*_{n,k}$.

3. Simulation Study

To compare the quality of the approximations derived under the two resampling schemes, several simulation studies were performed for different parametric models under the null hypothesis and under the alternative (see Dikta *et al.*[5]). Overall, under both resampling schemes the nominal level was reasonably well attained for the KS and for the CvM statistic, respectively. The empirical power of all the tests increased substantially with increasing sample size. Finally, the MB-based tests outperformed the CB-based ones in all the considered cases.

Figure 1 exemplarily illustrates the gain in power of the MB resampling scheme compared to the CB one for the CvM test. The figure shows the edf of 100 approximated p-values (that is, 100 different ODS) for the CvM statistic based on the MB and on the CB resampling scheme. The sample size of each ODS is $n = 100$ and Z is normally distributed with zero mean and variance 4. For the parametric model,

$$\mathcal{M} = \{m : m(z) = 1 - \exp(-\exp(\alpha + \beta z)),\ \alpha, \beta \in \mathbb{R}\}$$

was assumed while the data were generated according to

$$m(z) = m(z) = 1 - \exp(-\exp(-0.2 - z - 0.225z^2)).$$

Each p-value was approximated by $N = 1000$ bootstrap replications. Obviously, the null hypothesis $H_0 : m \in \mathcal{M}$ is not correct.

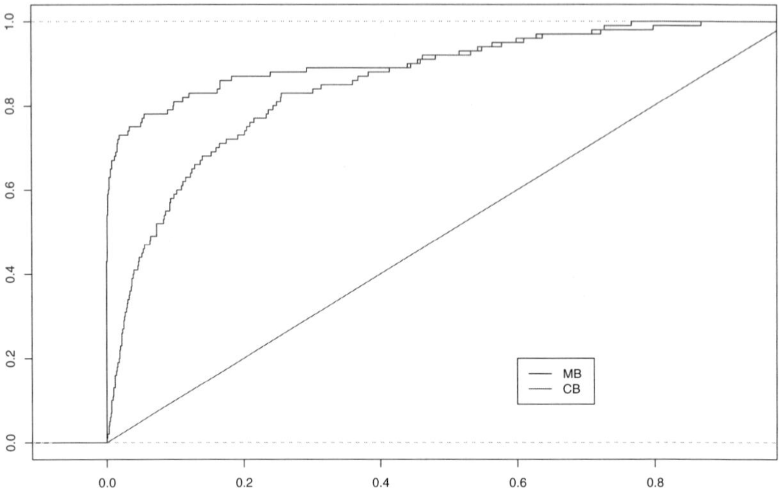

Fig. 1. Empirical d.f. of 100 p-values of the CvM tests.

4. Application to Survival Analysis

Assume that our ODS represents the results of a survival study, where the observations are generated according to RCM. If RCM is the only assumption which can be made for our data, the famous Kaplan-Meier or product-limit estimator, see Kaplan-Meier,[6] is the first choice for an estimator of F, the d.f. of the lifetime. However, if a parametric regression model for m can be assumed for the binary data, it is shown in Dikta[7] and Dikta *et al.*[8] that the Kaplan-Meier estimator and integral estimators based on it can be improved by a semiparametric estimator of F. But this improvement is only guaranteed, if the correct regression model for the binary data is used.

In the following, we consider two parametric models for m:

- **Koziol-Green Model[9] (KG):**

$$\mathcal{M}_{KG} = \left\{m(z,\theta) = \theta \,:\, 0 < \theta < 1\right\}$$

- **Generalized Koziol-Green Model[7] (GKG):**

$$\mathcal{M}_{GKG} = \left\{m(z,\theta) = \frac{\theta_1}{\theta_1 + z^{\theta_2}} \,:\, \theta_1 > 0, \theta_2 \in \mathbb{R}\right\}$$

The bootstrap tests are now applied to check whether the following data sets fits to the KG or the GKG model.

- CHD: Channing House data, Hyde.[10] Survival times of $n = 97$ men in a Palo Alto retirement center. 46 non-censored cases.
- HTD: Stanford heart transplant data, Miller and Halpern.[11] Survival times of $n = 184$ patients having received heart transplant. 113 non-censored cases.
- OTD: Oestrogen treatment data, Hollander and Proschan.[12] Survival times of $n = 211$ prostate cancer patients treated with oestrogen, 90 non-censored cases.

Table 1. $p-$values of KG and GKG in MB and CB-based tests, 1000 replications.

	Null	MB		CB	
Data	Hypothesis	CvM	KS	CvM	KS
CHD	KG	0.421	0.427	0.626	0.417
	GKG	0.578	0.637	0.754	0.594
HTD	KG	0.000	0.000	0.000	0.000
	GKG	0.494	0.769	0.618	0.821
OTD	KG	0.004	0.002	0.000	0.000
	GKG	0.016	0.016	0.007	0.000

The results of these tests are given in Table 1. The p-values show, that the CHD do not depart from the KG model. For the HTD set, the data depart significantly from the KG model while the GKG model fits. In the case of the OTD, neither KG nor GKG can be used in a semiparametric approach. The results for the KG model are in line with those of Csörgő[13] and Henze[14] for the considered data sets. Both authors introduced goodness-of-fit tests to check for KG and applied them to the considered data sets.

5. Conclusion

In summary, our analysis has shown that

- MB and CB tests hold the significance level under H_0.
- MB tests outperformed those which are based on CB drastically.
- The heterogeneous structure of the conditional variance is preserved in the case of MB resampling but not when CB is used.

- MB and CB tests show similar results as the specialized test of Csörgő[13] and Henze[14] when they are applied to some real data to check for the Koziol-Green model.

References

1. W. Stute, Nonparametric model checks for regression. *Ann. Statist.* **25** (1997), 613-641.
2. W. Stute, W. González Manteiga, and M. Presedo Quindimil, Bootstrap approximations in model checks for regression. *J. Amer. Statist. Assoc.* **93** (1998), 141-149.
3. L. X. Zhu, K. C. Yuen, and N. Y. Tang, Resampling methods for testing semiparametric random censorship model. *Scand. J. Statist.* **29** (2002), 111-123.
4. B. Efron and R. J. Tibshirani, *An introduction to the bootstrap.* (Chapman & Hall, 1993).
5. G. Dikta, M., Kvesic, and C. Schmidt, Bootstrap approximations in model checks for binary data. *J. Amer. Statist. Assoc.* **101** (2006), 521-530.
6. E. L. Kaplan and P. Meier, Nonparametric estimation from incomplete observations. *J. Amer. Statist. Assoc.* **53** (1958), 457-481.
7. G. Dikta, On semiparametric random censorship models. *J. Statist. Plann. Inference* **66** (1998), 253-279.
8. G. Dikta, J. Ghorai, and C. Schmidt, The central limit theorem under semiparametric random censorship models. *J. Statist. Plann. Inference* **127** (2005), 23-51.
9. J. A. Koziol and S. B. Green, A Cramér-von Mises statistic for randomly censored data, *Biometrika*, **63** (1976), 465-474.
10. J. Hyde, Testing survival under right censoring and left truncation, *Biometrika*, **64** (1977), 225-230.
11. R. Miller and J. Halpern, Regression with censored data. *Biometrika* **69** (1982), 521-531.
12. M. Hollander and F. Proschan, Testing to determine the underlying distribution using randomly censored data. *Biometrika* **35** (1979), 393-401.
13. S. Csörgő, Testing for the proportional hazards model of random censorship. *Proc. 4th Prague Symp. Asympt. Statist. Charles Univ. Prague.* P. Mandl and M. Hušková (eds.), (Prague, 1988).
14. N. Henze, A quick omnibus test for the proportional hazards model of random censorship. *Statistics* **24** (1993), 253-263.

NONREFLECTING LOCAL BOUNDARY CONDITIONS FOR ELLIPTICAL-SHAPED EXTERIOR BOUNDARIES

H. BARUCQ

INRIA Bordeaux-Sud Ouest Research Center, Team-Project Magique 3D, Laboratoire de Mathématiques Appliquées, CNRS UMR 5142, Université de Pau et des Pays de l'Adour, Pau, FRANCE
E-mail: helene.barucq@univ-pau.fr

R. DJELLOULI*

Department of Mathematics, California State University Northridge Northridge, CA 91330-8313, USA
** E-mail: rabia.djellouli@csun.edu*

A. SAINT-GUIRONS

Laboratoire de Mathématiques Appliquées, CNRS UMR 5142, Université de Pau et des Pays de l'Adour, INRIA Bordeaux-Sud Ouest Research Center, Team-Project Magique 3D, Pau, FRANCE
E-mail: anne-gaelle.saint-guirons@univ-pau.fr

A new class of nonreflecting boundary conditions is proposed for solving exterior Helmholtz problems. These boundary conditions are applicable to exterior elliptical-shaped boundaries that are more suitable in terms of cost-effectiveness for surrounding elongated scatterers and radiators. This class of conditions distinguishes itself from similar absorbing boundary conditions by (a) the new conditions are exact for the first radiating modes, (b) they are easy to implement and to parallelize, and (c) they are compatible with the local structure of the computational finite element scheme. The analysis of the performance of these conditions shows that, in the low frequency regime, the new second order condition retains a good level of accuracy *regardless* of the slenderness of the artificial boundary.

Keywords: Local absorbing boundary conditions, elliptical-shaped artificial boundaries, Dirichlet-to-Neumann operator, acoustic scattering problems.

1. Introduction

It is well known that the finite element computation of the solutions of exterior Helmholtz problems requires first to reformulate them in a finite domain. This is often achieved by surrounding the given scatterer(s) by an artificial *exterior* boundary that is located at some distance (measured in multiples of wavelength of interest) from its surface. A so-called "nonreflecting" boundary condition is then prescribed on the artificial boundary to represent the "far-field" behavior of the scattered field. The challenge here is the development of a simple but reliable as well as cost-effective computational procedure for representing the far-field behavior of the scattered (see, e.g., the recent review by Turkel in the book[21]).

We propose in this work a new class of absorbing boundary conditions that are obtained from local approximations of the exact DtN boundary condition when expressed in elliptical coordinates.[10,11] This family of boundary conditions is designed to be employed on elliptical-shaped boundaries (in 2D), and on prolate spheroid boundaries (in 3D) since such boundaries are primary candidates for surrounding elongated scatterers. The idea for constructing such conditions is driven by several considerations chief among them the following two reasons. First, the widely-used second order absorbing boundary condition (BGT2) designed by Bayliss, Gunzburger and Turkel for spherical-shaped boundaries[5] performs poorly when it is expressed in elliptical coordinates and applied to prolate spheroid boundaries in the low frequency regime.[16] The accuracy deteriorates significantly for large eccentricity values of the boundaries, i.e. elongated boundaries, as observed in.[16] The damping effect introduced to this condition[17] improves the performance for small eccentricity values. However, the modified BGT2 still performs poorly for eccentricity values larger than 0.6 in the (relatively) low frequency regime (see Figure 15 in[17]). Hence, there is a need for constructing local absorbing boundary conditions (ABC) that extend the range of satisfactory performance. Second, the three-dimensional approximate local DtN conditions designed for spherical-shaped boundaries[12] perform very well for low wavenumber values as reported in.[13] We recall that, in $\mathbb{R}^3$, the second-order approximate local DtN boundary condition and the BGT2 condition are identical.[13] Nevertheless, using these conditions on spherical-shaped exterior boundaries when solving scattering problems by elongated scatterer often leads to larger than needed computational domains, which hampers computational efficiency. This suggests that approximate local DtN boundary conditions designed for prolate spheroidal–shaped boundaries is an attractive alternative for improving the computational performance.

Given that, this work is devoted to the construction of theses conditions and to the assessment of their performance when employed on prolate spheroidal-shaped boundaries for solving three-dimensional radiator and scattering problems.

Because of space limitations, we consider in this paper *only* the case of the *second-order* approximate local DtN boundary condition (DtN2) for three-dimensional problems, and we assess mathematically and numerically its performance. More specifically, we analyze the effect of low wavenumber and the eccentricity on the performance of this condition in the case of three-dimensional acoustic scattering problems. We adopt the on-surface radiation condition formulation (OSRC)[15] in order to perform *analytically* this investigation. We note that such formulation is *not* appropriate for high frequency regime as observed previously in.[2] The main interest in the following analyses is to evaluate the performance of the proposed DtN2 boundary condition at low wavenumber to see if relatively small computational domains can be employed in order to avoid excessive computational cost. The OSRC formulation must be viewed as an extreme case while an exterior ellipsoidal-shaped boundary surrounding an elongated scatterer would be less "demanding" on the boundary condition. Results for radiating problems can be found in.[4,18] The analysis in the case of two-dimensional problems (radiating and scattering) can be found in.[3,18]

2. Preliminaries

Throughout this paper, we shall use the prolate spheroidal coordinates (ξ, φ, θ) which are related to the cartesian coordinates $(x, y, z) = (b \sin\varphi \cos\theta, b \sin\varphi \sin\theta, a \cos\varphi)$ where $\varphi \in [0, \pi)$, $\theta \in [0, 2\pi)$. The parameters $a = f \cosh\xi$ and $b = f \sinh\xi$ respectively represent the major and the minor axes, where ξ is a strictly positive real number and $f = \sqrt{a^2 - b^2}$ is the interfocal distance. We also define the eccentricity e on a prolate spheroid at $\xi = \xi_0$, by:

$$e = \frac{1}{\cosh\xi_0} = \sqrt{1 - \frac{b^2}{a^2}} \tag{1}$$

The eccentricity e characterizes the slenderness of the surface. It satisfies $0 < e < 1$. Note that when $e \to 0$, the prolate spheroid degenerates into a sphere and when $e \to 1$, the spheroid degenerates into a line with length $2f$ on the z-axis. Furthermore, any incident plane wave u^{inc} can be expressed in this coordinate system as follows:

$$u^{\text{inc}} = e^{ikf \cosh\xi(\cos\varphi \cos\varphi_0 + \tanh\xi \sin\varphi \sin\varphi_0 \cos\theta)} \tag{2}$$

where the wavenumber k is a positive number and φ_0 is the incident angle. We consider in this work three-dimensional acoustic scattering problems by prolate-spheroid scatterers. Such restriction allows us however to adopt the OSRC formulation and set the ellipsoidal-shaped artificial boundary on top of the boundary of the obstacle. In addition, we assume, for simplicity, that the scatterers are *sound-soft*. Consequently, the acoustic scattered field u^{scat} solution of a sound-soft prolate-spheroid scatterer can then be expressed in terms of spheroidal wave functions as follows:[19]

$$u^{\text{scat}} = -2\sum_{m=0}^{\infty}\sum_{n=m}^{\infty}(2-\delta_{0m})A_{mn}\mathrm{i}^{n}R_{mn}^{(3)}(kf,\cosh\xi)S_{mn}(kf,\cos\varphi)\cos m\theta \tag{3}$$

where

$$A_{mn} = \frac{R_{mn}^{(1)}(eka,e^{-1})S_{mn}(eka,\cos\varphi_0)}{N_{mn}R_{mn}^{(3)}(eka,e^{-1})} \tag{4}$$

$R_{mn}^{(j)}$ represents the spheroidal wave functions of the jth type,[6] S_{mn} denotes the angular wave functions,[6] N_{mn} represents the normalization factor of the angular spheroidal wave functions,[6] and δ_{0m} is the Kronecker delta symbol. Moreover, we recall that the mnth prolate spheroidal radiating mode is given by:[19]

$$u_{mn} = R_{mn}^{(3)}(kf,\cosh\xi)S_{mn}(kf,\cos\varphi)\cos m\theta \tag{5}$$

3. Three-dimensional second-order approximate *l*ocal DtN boundary condition in prolate spheroid coordinates

The three-dimensional second-order (DtN2) local Dirichlet-to-Neumann boundary conditions, defined on the prolate-spheroidal boundary $\xi = \xi_0$, is given by:

$$\begin{aligned}\text{DtN2}:\quad \frac{\partial u}{\partial \xi} = \frac{\sqrt{1-e^2}}{(\lambda_{01}-\lambda_{00})\,e}&[(\lambda_{01}R_{01}-\lambda_{00}R_{00}\\ &-(R_{00}-R_{01})(eka)^2\cos^2\varphi)u+(R_{00}-R_{01})\Delta_\Gamma u]\end{aligned} \tag{6}$$

where λ_{mn} represents the spheroidal eigenvalues[6] and the coefficients R_{mn} depend on the radial spheroidal wave functions of the third kind $R_{mn}^{(3)}$, the wavenumber ka, and the eccentricity e as follows:

$$R_{mn} = \frac{R_{mn}^{\prime(3)}(eka,e^{-1})}{R_{mn}^{(3)}(eka,e^{-1})} \tag{7}$$

and Δ_Γ denotes the Laplace Beltrami operator, which reads in prolate spheroidal coordinates (ξ, φ, θ) as:

$$\Delta_\Gamma = \frac{1}{\sin\varphi}\frac{\partial}{\partial\varphi}\left(\sin\varphi\frac{\partial}{\partial\varphi}\right) + \frac{1}{\sin^2\varphi}\frac{\partial^2}{\partial\theta^2} \tag{8}$$

The following two remarks are noteworthy:

- The construction methodology we propose for deriving the class of approximate local DtN boundary conditions in prolate spheroidal coordinates can be viewed as an *inverse-type* approach. More specifically, we start from a Robin-type boundary condition with unknown coefficients. Hence, in the case of DtN2 condition, we set:

$$\frac{\partial u}{\partial \xi} = A\,u + B\,(\Delta_\Gamma - (eka)^2\cos^2\varphi)u \tag{9}$$

where A and B are constant (independent of φ) to be determined. Note that, unlike DtN2 boundary condition for the spherical-shaped boundaries, the coefficients of this condition depend on the angular variable φ. Such dependence is necessary for constructing a *symmetric* boundary condition. Then, we observe that all radiating modes u_{mn} given by Eq. (5) satisfy:

$$\Delta_\Gamma u_{mn} = \left(-\lambda_{mn} + (eka)^2\cos^2\varphi\right)u_{mn} \tag{10}$$

Hence, in order to determine the constants A and B, we assume that, at $\xi = \xi_0$, we have:

$$\frac{\partial u_{mn}}{\partial \xi} = A\,u_{mn} + B\,\left(\Delta_\Gamma - (eka)^2\cos^2\varphi\right)u_{mn}\;; m = 0 \text{ and } n = 0, 1 \tag{11}$$

Then, using Eq. (10), it follows that (A, B) is the unique solution of the following 2×2 linear system:

$$\begin{cases} A - B\,\lambda_{00} = \dfrac{\sqrt{1-e^2}}{e}R_{00} \\ \\ A - B\,\lambda_{01} = \dfrac{\sqrt{1-e^2}}{e}R_{01} \end{cases} \tag{12}$$

where the coefficients R_{mn} are given by Eq. (7).
The DtN2 boundary condition given by Eq. (6) is a direct consequence of solving the system (12) and substituting the expressions of (A, B) into Eq. (9).

- The boundary condition (6) is called *local* DtN condition because it results from a localization process of the truncated global DtN boundary operator defined in.[10,11] The local feature of this condition is of a great interest from a numerical view point. Indeed, the incorporation of this condition in any finite element code introduces only mass- and stiffness-type matrices defined on the exterior boundary. The coefficients λ_{mn} and R_{mn} can be computed once for all at the preprocessing level. Furthermore, when $e = 0$ (the prolate spheroid becomes a sphere), condition(6) is identical to the second-order approximate *local* DtN condition designed for spherical shaped boundaries (see Eq. (13) in[13]). This property can be easily established using the asymptotic behavior of the radial spheroidal wave functions of the third kind $R^{(3)}_{mn}$ and the prolate spheroidal eigenvalues λ_{mn}.

4. Performance analysis

In the following, we analyze the effect of low wavenumber ka and the eccentricity e on the performance of DtN2 in the case of sound-soft scattering problems. We adopt the on-surface radiation condition formulation (OSRC)[15] in order to perform this investigation analytically. As in,[13,16,17] we assess the performance of the absorbing boundary condition DtN2 using the *specific impedance* introduced in[7,8] as a practical tool for measuring the efficiency of ABCs in the context of the OSRC formulation. This *non-dimensional* quantity measures the effect of the truncated medium in physical terms. In the elliptical coordinates system, the specific impedance can be expressed as follows:

$$Z = \frac{\mathrm{i}\sqrt{1-e^2}ka\, u^{\mathrm{scat}}}{\frac{\partial}{\partial\xi}\left(u^{\mathrm{scat}}\right)\big|_{\xi=\xi_0}} \tag{13}$$

Therefore, the specific impedance Z^{exact} corresponding to the exact solution for three-dimensional sound-soft acoustic scattering problems can be computed analytically using $u^{\mathrm{scat}} = -u^{\mathrm{inc}}$ (see Eq. (2)), and the Fourier series given by Eq. (3) to evaluate $\frac{\partial}{\partial\xi}\left(u^{\mathrm{scat}}\right)|_{\xi=\xi_0}$ (for more details, see[16–18]). One can also verify[4,18] that the approximate specific impedance (Z^{DtN2}) corresponding to the second order DtN boundary condition (DtN2) is given by:

$$Z^{\mathrm{DtN2}} = \frac{\lambda_{01} - \lambda_{00}}{\left(\lambda_{01}R_{00} - \lambda_{00}R_{01}\right)\left(-2i\alpha ka - (ka)^2\,\tau - (eka)^2\cos\varphi\right)} \tag{14}$$

where

$$\alpha = \cos\varphi\cos\varphi_0 + \sqrt{1-e^2}\sin\varphi\sin\varphi_0\cos\theta, \quad \tau = \left(\frac{\partial\alpha}{\partial\varphi}\right)^2 + \frac{1}{\sin^2\varphi}\left(\frac{\partial\alpha}{\partial\theta}\right)^2 \tag{15}$$

and R_{00} and R_{01} are given by Eq. (7).

4.1. *Mathematical results*

Next, we analyze the asymptotic behavior of the DtN2 specific impedance as $ka \to 0$. We first recall (see Eq. (91), p. 3655 in[16]), that the asymptotic behavior of the exact specific impedance Z^{ex3} of the scattered field on the surface of a prolate spheroid as $ka \to 0$ is given by:

$$Z^{\text{ex3}} \sim (ka)^2 - ika \tag{16}$$

Proposition 4.1. *The asymptotic behavior of the approximate specific impedance* Z^{DtN2} *as* $ka \to 0$ *is given by:*

$$Z^{\text{DtN2}} \sim (1-\alpha)(ka)^2 - ika \tag{17}$$

where $\alpha = \cos\varphi\cos\varphi_0 + \sqrt{1-e^2}\sin\varphi\sin\varphi_0\cos\theta$.

Proof of Proposition 4.1. This proof is detailed in.[4,18]

Remark 4.2. Eq. (17) indicates that the asymptotic behavior of Z^{DtN2} depends on the eccentricity as well as on the observation angle φ. This dependence is comparable to the asymptotic behavior of the BGT2 approximate specific impedance (see Eq. (93), p. 3657 in[16]). Last, when $e = 0$, the asymptotic behavior of Z^{DtN2} is identical to the case of spheres (see Eq. (140) p. 47 in[13]).

4.2. *Illustrative numerical results*

We have performed several experiments to investigate numerically the effect of the wavenumber and the slenderness of the boundary on the performance of the DtN2 boundary condition given by Eq. (6) when solving sound-soft scattering problems in the OSRC context. We have evaluated the relative error $\dfrac{\|Z^{\text{ex3}} - Z^{\text{DtN2}}\|_2}{\|Z^{\text{ex3}}\|_2}$. We have compared the results to the ones obtained with BGT2 condition (see Eq. (60) p. 3645 in[16]) when applied on prolate spheroid boundaries (see Figs. (22) to (33) in[16]). All numerical results can be found in.[3,4,18] For illustration purpose and because of space limitations,

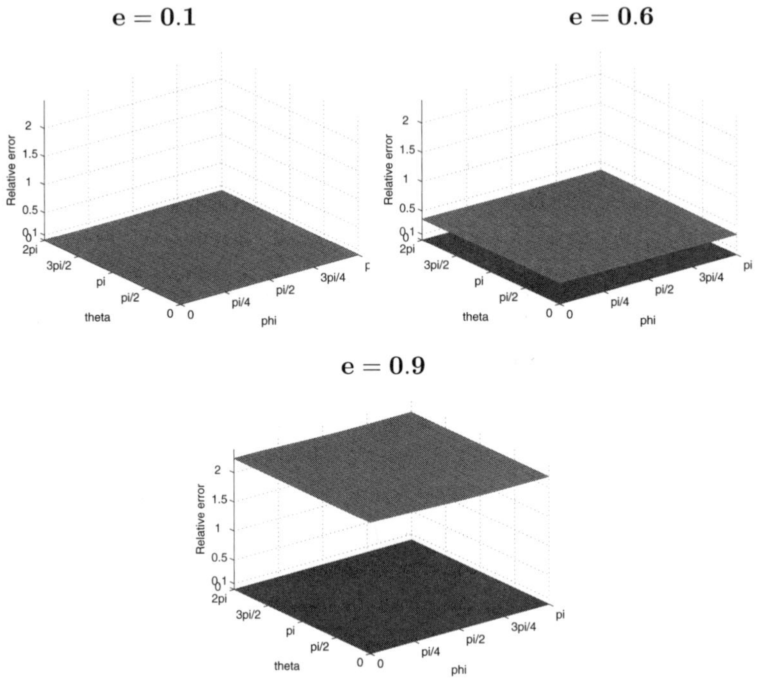

Fig. 1. Relative error of the specific impedance corresponding to DtN2 (black) and BGT2 (dark grey) when solving three-dimensional sound-soft scattering problem with $ka = 0.1$ and incident angle $\varphi_0 = 0$.

we present the results for only one (low) value of the wavenumber, $ka = 0.1$ corresponding to one value of the incidence angle $\varphi_0 = 0$. These results have been obtained for six eccentricity values $e = 0.1$ corresponding to a prolate spheroid "close" to a sphere, $e = 0.6$ corresponding to a "regular" prolate spheroid boundary, and $e = 0.9$ corresponding to a "very" elongated prolate spheroid. Note that since all the approximate specific impedances depend on the observation angles $\varphi \in [0, \pi)$ and $\theta \in [0, 2\pi)$, we have reported in Fig. (1) the relative error as a function of (φ, θ). The following two observations are noteworthy. First, the DtN2 absorbing boundary condition retains an excellent level of accuracy when solving acoustic problems for low wavenumbers (the relative error is below 2% for all eccentricity values). In addition, the results depicted in Fig. (1) clearly demonstrate that such good performance is not sensitive to the value of the eccentricity e. These results suggest in particularthat DtN2 absorbing boundary condition given by Eq. (6) is appropriate for elongated boundaries. Second, the DtN2 absorbing boundary condition clearly outperforms the second-order BGT2 absorbing boundary condition especially for high eccentricity values. Indeed, there is a significant loss of accuracy for the BGT2 boundary condition when $e \geq 0.6$ (the relative error is larger than 40%). This demonstrates that DtN2 absorbing boundary condition extends the range of satisfactory performance to all eccentricity values in the low frequency regime.

5. Conclusion

We have designed a new class of approximate local ABCs to be applied on ellipsoidal-shaped exterior boundaries when solving acoustic scattering problems by elongated obstacles. These conditions are *exact* for the first radiation modes, they are easy to implement and to parallelize, and they preserve the local structure of the computational finite element scheme. The analysis reveals that in the case of the acoustic scattering problem the DtN2 boundary condition delivers an excellent level of accuracy (the relative error $\leq 2\%$) in the low frequency regime for *all* eccentricity values, while the BGT2 boundary condition performs poorly for eccentricity values $e \geq 0.6$ (the relative error $\geq 40\%$). Hence, the new second-order approximate local DtN boundary condition (DtN2) extends the range of satisfactory performance to all eccentricity values.

Acknowledgments

The authors acknowledge the support by INRIA/CSUN Associate Team Program. Any opinions, findings, conclusions or recommendations expressed in this material are those of the authors and do not necessarily reflect the views of INRIA or CSUN.

References

1. M. Abramovitz, I. Stegun, Handbook of Mathematical Functions with Formulas, Graphs and Mathematical Tables, Dover Publications, New York, 1972
2. X. Antoine, Fast approximate computation of a time-harmonic scattered field using the on-surface radiation condition method, IMA J. Appl. Math., 66(1):83–110, 2001
3. H. Barucq, R. Djellouli, A. Saint-Guirons, Construction and performance analysis of *local* DtN absorbing boundary conditions for exterior Helmholtz problems. Part I : Elliptical shaped boundaries, NRIA Research Report, No. 6394 (2007). Available online at: http://hal.inria.fr/inria-00180471/fr/

4. H. Barucq, R. Djellouli, A. Saint-Guirons, Construction and performance analysis of *local* DtN absorbing boundary conditions for exterior Helmholtz problems. Part II : Prolate spheroid boundaries,INRIA Research Report, No.6395 (2007). Available online at: http://hal.inria.fr/inria-00180475/fr/
5. A. Bayliss, M. Gunzberger, and E. Turkel, Boundary conditions for the numerical solution of elliptic equations in exterior regions, SIAM J. Appl. Math., 42 (2), pp. 430-451, 1982
6. C. Flammer, Spheroidal Functions, Standford University Press, Standford, CA, 1957
7. T. L. Geers, Doubly asymptotic approximations for transient motions of submerged structures, J. Acoust. Soc. Am., 64 (5), pp. 1500-1508, 1978
8. T. L. Geers, Third-order doubly asymptotic approximations for computational acoustics, J. Comput. Acoust., 8 (1), pp. 101-120, 2000
9. D. Givoli and J. B. Keller, Nonreflecting boundary conditions for elastic waves, Wave Motion, 12(3):261–279, 1990
10. D. Givoli, Exact representations on artificial interfaces and applications in mechanics, AMR, 52(11):333–349, 1999
11. M. J. Grote, J. B. Keller, On nonreflecting boundary conditions, J. Comput. Phys., 122, 2, 231-243, 1995
12. I. Harari and T. J. R. Hughes, Analysis of continuous formulations underlying the computation of time-harmonic acoustics in exterior domains, Comput. Methods Appl. Mech. Engrg., 97(1):103–124, 1992
13. I. Harari, R. Djellouli, Analytical study of the effect of wave number on the performance of local absorbing boundary conditions for acoustic scattering, Applied Numerical Mathematics, 50, 15-47, 2004
14. J. B. Keller, D. Givoli, Exact nonreflecting boundary conditions, J. Comput. Phys., 82 (1), 172-192,1989
15. G. A. Kriegsmann, A. Taflove, and K. R. Umashankar, A new formulation of electromagnetic wave scattering using an on-surface radiation boundarycondition approach, IEEE Trans. Antennas and Propagation 35 (2), 153–161, 1987
16. R.C. Reiner, R. Djellouli, and I. Harari, The performance of local absorbing boundary conditions for acoustic scattering from elliptical shapes, Comput. Methods Appl. Mech. Engrg, 195, 3622-3665, 2006
17. R.C. Reiner and R. Djellouli, Improvement of the performance of the BGT2 condition for low frequency acoustic scattering problems, Journal of Wave Motion, 43, pp. 406-424, 2006.
18. A. Saint-Guirons, Construction et analyse de conditions aux limites absorbantes pour des problèmes de propagation d'ondes, Ph.D. thesis, (In Preparation).
19. T. B. A. Senior, Scalar diffraction by a prolate spheroid at low frequencies, Canad. J. Phys. 38 (7) (1960) 1632-1641
20. J.A. Stratton, Electromagnetic theory, McGraw-Hill, New York, 1941
21. E. Turkel, Iterative methods for the exterior Helmholtz equation including absorbing boundary, conditions, In: Computational Methods for Acoustics Problems, F. Magoulès (ed.), Saxe-Coburg Publications, 2008

MODELING AND ANALYSIS OF AXONOGENESIS: RANDOM SPATIAL NETWORK PERSPECTIVE

YANTHE E. PEARSON* and DONALD A. DREW

Department of Mathematical Sciences, Rensselaer Polytechnic Institute, Troy, New York 12180, USA
** E-mail: pearsy@rpi.edu*
www.rpi.edu

EMILIO CASTRONOVO

E-mail: emilio.castrono@gmail.com

Data from time lapse microscopy of live embryonic rat hippocampal neurons growing in cell culture are used to study the dynamics of axonal growth.[1] We analyze axonal trajectory data based on cells growing in a homogeneous medium. Due to the noisy nature of the data we develop filtering algorithms to smoothen out the paths while maintaining the underlying dynamics of the axon growth process. We analyze the new paths and propose a model for growth cone kinematics during axonogenesis without a gradient field. In this work we present a simple renewal process with the aim of reproducing certain path behaviors of the growth cone. Future development will include, renewal process simulation, and gradients effects.

Keywords: Random spatial networks, Axonogenesis, Stochastic differential equations

Introduction

The growth and guidance of axons in the nervous system is a key event in brain development and during axon regeneration after injury. Neurons possess genetic programs that allow for certain aspects of axon growth and guidance, but normal development requires external cues such as secreted soluble factors and contact with adhesion molecules.[2] Neurons develop from nascent neuroblasts, ultimately extending axons and forming synapses on the dendrites of other neurons. This is a crucial step in development, since neural information is transmitted via these axon-dendrite synapses, resulting in sensory information, muscle actions, and cognition. The connections must be sufficiently spread out and dense for the organism to thrive.[3]

Axon growth is guided by a growth cone, a dynamic arrangement of microtubules and filopodia, with bridging membrane lamellipodia between them, and active actin nets all combining to grow the axon.[4] Signaling systems sense the surroundings and modify the assembly and collapse of the structures. The result is a highly dynamic system, where the axon grows, pauses, and retracts, while turning toward and away from signals. These motions are observed in controlled two-dimensional situations, with and without the presence of guidance cues.

To function normally, axonal networks do not depend on the exact location and orientation of each individual axon; however, certain statistical properties such as density and direction must be achieved. The mathematical objective of this work is to derive and evaluate a model of axon growth in a homogeneous environment as a "random spatial network", which is defined as a geometrical structure with randomly oriented thin line-like objects that are spatially and temporally controlled. Random network formation is analogous to a random walk, although the outcome of random walks is most often the spatial probability density function $P(x,t)$, without the directional dependence. The biological objective of the project is to identify parameters of axon growth including signaling molecules and the resulting behavior of filopodia and lamellipodia, and relate these parameters to the statistics of the network.

While we apply the ideas to the controlled situation of axon growth on the planar substrate, the concept of random spatial network relates to many different situations, including microtubules, capillaries, collagen nets, and trees.

Model Formulation

We describe axons as a set of curves proceeding from soma to synapse, wandering through the space that they occupy. The result describes axons location and direction. These networks are formed by axonogenesis, a process by which the tip of the axon (growth cone) guides its growth. This temporal process is modeled as a formation of a set of curves as they have grown in time. Thus, there are two aspects of axonal networks that we wish to exploit. One is the description of the statistical properties of a developed network. This includes the probability that a given point will be occupied by a curve, having a given direction, or more precisely, a probability density function $P(\mathbf{x}, \mathbf{k}, s)$ for position $\mathbf{x}$ with direction $\mathbf{k}$ at arclength s. The second description of axonal networks is how each curve progresses from its initiation point to its position at a given time t. For this description, we write stochastic differential equations for the positions of the ends of the curves, $\mathbf{x}(t)$ at time t.

Random Spatial Network – Spatial Description

First, the resulting network formed from the microscopic processes possesses a density in space, with a direction field. This network and its corresponding density function, are independent of the temporal processes that resulted in their formation. Each such fiber in the network can be described in terms of the arc length from its point of formation to its distal end.

pdf Description Suppose $P(\mathbf{x}, \mathbf{k}, s)\mathrm{d}\mathbf{x}\mathrm{d}\mathbf{k}$ is the probability of finding the point of the curve at arclength s from its initiation point within $\mathrm{d}\mathbf{x}$ of the point $\mathbf{x}$ with direction within $\mathrm{d}\mathbf{k}$ of the direction (unit vector) $\mathbf{k}$. We wish to derive an equation for $P(\mathbf{x}, \mathbf{k}, s)$ from a knowledge of the microscopic information about the curves. For concreteness, we shall restrict the discussion to two space dimensions, and denote the dependence on the direction $\mathbf{k} = [\cos\theta, \sin\theta]^T$ by writing $P(x, y, \theta, s)$ for the probability density function.

Suppose that the curves are generated by a process in which the curve consists of discrete links of length which can change direction from link-to-link by either 0 or an amount $\pm\mathrm{d}\theta$. In addition, a link may not continue with some probability of the curve ending. Thus we have:

$$\begin{aligned} P(x, y, \theta, s + \mathrm{d}s) &= P(x - \mathrm{d}s\cos\theta, y - \mathrm{d}s\sin\theta, \theta, s)(1 - K_{end} - K_{right} - K_{left}) \\ &\quad + P(x - \mathrm{d}s\cos(\theta - \mathrm{d}\theta), y - \mathrm{d}s\sin(\theta - \mathrm{d}\theta), \theta - \mathrm{d}\theta, s)K_{right} \\ &\quad + P(x - \mathrm{d}s\cos(\theta + \mathrm{d}\theta), y - \mathrm{d}s\sin(\theta + \mathrm{d}\theta, s), \theta + \mathrm{d}\theta)K_{left} \end{aligned}$$

expand in a Taylor series, correct to order $\mathrm{d}s$ and $\mathrm{d}\theta^2$, and use the following assumptions: K_{end} must be of order $\mathrm{d}s$; we write $K_{end} = k_{end}\mathrm{d}s$.

$(K_{left} - K_{right})\mathrm{d}\theta$ must be of order $\mathrm{d}s$; we write $(K_{left} - K_{right})\mathrm{d}\theta = k_{turn}\mathrm{d}s$.

$(K_{left} + K_{right})\mathrm{d}\theta^2$ must be of order $\mathrm{d}s$; we write $(K_{left} + K_{right})\mathrm{d}\theta^2 = 2D\mathrm{d}s$.
The partial differential equation is then

$$\frac{\partial P}{\partial s}(x,y,\theta,s) + \cos\theta\frac{\partial P}{\partial x}(x,y,\theta,s) + \sin\theta\frac{\partial P}{\partial y}(x,y,\theta,s)$$
$$= k_{turn}\frac{\partial P}{\partial \theta}(x,y,\theta,s) + D\frac{\partial^2 P}{\partial \theta^2}(x,y,\theta,s) - k_{end}P(x,y,\theta,s)$$

Trajectory sde Description Each fiber in the network can be described by specifying how the fiber progresses from a point (x, y) to a point distance $\mathrm{d}s$ away. To accomplish this, we use an (sde) system, where the independent variable is the arc length.

$$\begin{aligned} \mathrm{d}x(s) &= \cos\theta(s)\mathrm{d}s \\ \mathrm{d}y(s) &= \sin\theta(s)\mathrm{d}s \\ \mathrm{d}\theta(s) &= f(x,y,\theta,s)\mathrm{d}s + \sqrt{2\sigma_\theta^2}\mathrm{d}W(s) \end{aligned} \tag{1}$$

Here $f(x, y, \theta, s)$ is the deterministic steering of the axon, and $\sqrt{2\sigma_\theta^2}\mathrm{d}W(s)$ is a Wiener process, representing the randomness in the axon steering. We shall not pursue the steered case here, so that we assume $f(x, y, \theta, s) = 0$. The *sde model* formulation assumes that the axon trajectory can be modeled by an ODE system in arclength s with a noise term in the direction. From such a description the macroscopic pdf for the structure can be ascertained.

Random Spatial Network–Growth Description

Biologically, the random spatial network is formed using microscopic rules. For axon growth, an assembly called the growth cone drives this formation, and is capable of responding to molecules that either attract or repel it. During the formation, the growth cone is capable of pausing, then resuming growth, or reversing direction. As the growth cone moves back toward the soma, it leaves the axonal membrane behind, and subsequently follows the hollow axon shell back to the furthest extent, where it resumes forward growth. Clearly, then, growth and retraction are very different events, and require a different description. Suppose $(x(t), y(t))$ are the spatial coordinates of the tip of the axon at time t. $\theta(t)$ is the angle, $v_s(t)$ is the velocity, σ_t is the temporal diffusion coefficient, and $W_\theta(t)$ is the Weiner process in time.

$$\begin{aligned} \mathrm{d}x(t) &= v_s(t)\cos\theta(t)\mathrm{d}t \\ \mathrm{d}y(t) &= v_s(t)\sin\theta(t)\mathrm{d}t \\ \mathrm{d}\theta(t) &= \sqrt{2\sigma_t^2}\mathrm{d}W_\theta(t) \end{aligned} \tag{2}$$

$v_s(t)$ is the (signed) magnitude of the velocity of propagation of the axon tip. The similarity between this system (2), and the trajectory description (1) is apparent. However, the association $ds = v_s(t)dt$ only makes sense when $v_s(t) > 0$. We expect $v_s(t)$ to be equal to zero during the paused phase, and to be positive or negative during growth or retraction, respectively.

Markov Chain Approach We are interested in describing a version of model (2) aimed at describing the evolution of the growth cone following the final path of the axon. The position of the growth cone is given by $(x(t), y(t))$ thus identifying the growth cone along the path of the axon parameterized by the time t. Each frame of the digital image corresponds to a time t_i. Following the random curve model these coordinates are modeled as having progressed to arc length s_i from the initial point. This gives

$$x(s_i) = x(s_{i-1}) + v(s_{i-1})\cos(\theta(s_{i-1}))\Delta t_i \tag{3}$$

$$y(s_i) = y(s_{i-1}) + v(s_{i-1})\sin(\theta(s_{i-1}))\Delta t_i \tag{4}$$

where $\Delta t_i = t_i - t_{i-1}$ and $v(s)$ and $\theta(s)$ are stochastic random processes that we will characterize at due time. We define the distance traveled by the growth cone on the axon up to time frame t_n to be the progress variable $S(s_n) = \sum_{i=0}^{n} \delta S(s_i)$ where

$$\delta S(s_i) = \pm\sqrt{(x_{s_i} - x_{s_{i-1}})^2 + (y_{s_i} - y_{s_{i-1}})^2}$$

which is the distance covered on the axon by the growth cone between two consecutive time frames. We can then identify three states for $\delta S(s)$:

1) Forward Motion: $\delta S(s) > 0$ growth cone moves closer to the tip of the axon.
2) Paused Motion: $\delta S(s) = 0$ growth cone is stopped and seeking a new direction of movement.
3) Backward Motion: $\delta S(s) < 0$ growth cone traces back through its previous path.

Based on definition (3) and model (3,4) we see that $\delta S(s_i) = \pm v(s_i)\Delta t_i$. We define $\tilde{v}(s_i) = \pm v(s_i)$ to be a stochastic process that fully describes the evolution of the progress variable $S(s)$. This implies that the angle variability captured by $\theta(s)$ is represented exclusively by the sign of the random process $\tilde{v}$. We thus need to specify the probability distribution of $\tilde{v}$ and its correlation statistics.

Markovian Assumption Consider the probability that the growth cone evolution at the ith frame is going to be described by velocity $\tilde{v}(s_i)$. We assume that $\tilde{v}(s)$ to be a Markovian process, and define a Markov Matrix describing the jump probability between stages in which

- $v_i > 0$: implies forward motion,
- $v_i < 0$: implies backward motion, and
- $v_i = 0$: corresponds to a stopped state.

Experimental Data Collection

Neuron cultures grown in vitro are described by Goslin *et al.*, (1998). Digital images are captured every 2.5 min. over a 24 hour period.[1] We extract information on the axon dynamics in terms of (x, y) coordinates of the end of the axon just to the soma side of the growth cone. There are around 200–400 frames (data points) per cell. Below is a figure of phase contrast images of one axon growing in a homogeneous environment. This shows the alternating extension and retraction of the axon and its frequent changes in direction.

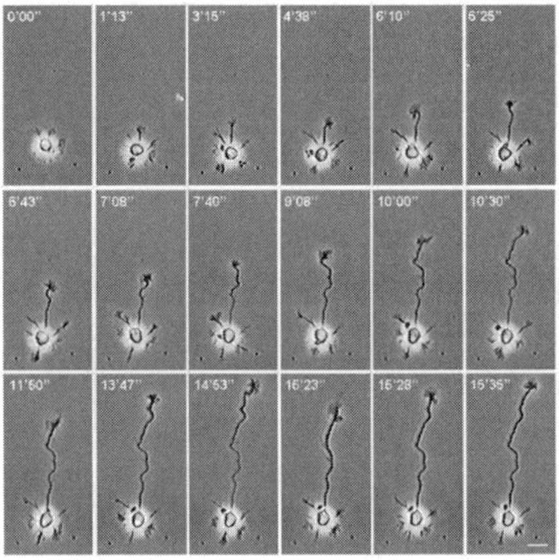

Fig. 1. 18 Frames of Growing Axon

Data Analysis and Filtering Algorithm The data show randomness in the direction of axon growth as well as observational noise. We filter the data to obtain piecewise line segment approximations to the curves representing the final positions of the axon. We use the angle changes between the line segments to quantify the randomness in direction. We then associate each data point with a corresponding point on this final trajectory. We extract a time sequence of the distance progressed along these final trajectories. This allows us to measure the forward motion, pausing, and retraction, and to estimate the frequencies of transition between the states. One goal of the filtering algorithm is to be able to discern between physical behavior of the growth cone and artifacts due to pixilation. We filter the trajectory data by removing repeats and backtracks, since the ultimate trajectory does not contain these features. For backtracking, we eliminate points where the absolute change in angle exceeds a given amount. For the present calculations, we assume that $\Delta\theta_{MAX}$ is $\frac{2\pi}{3}$. We iterate this process until no repetitions or backtracking occurs.

Progress Variable Next we define a variable that approximates the arc length along the trajectory up to each raw (i.e., unfiltered) data point. To do this, we project each raw point onto the closest filtered line segment. The progress variable is the sum of the lengths of the line segments closer to the initiation point, plus the length of the part of the nearest line segment up to the projected point.

Results

Angle Statistics The next six figures show how we go from "raw" to "filtered" to "projected raw" points and the distribution of *angle variability*, along with a fit to a normal distribution with parameters $\mu_{\eta_\theta} = -0.741$ and $\sigma^2_{\eta_\theta} = 26.7°$.

Markov Model From the progress variable, we extract the velocity statistics and the Markov variables for the stopped, growth, and retraction states. We obtain:

State	s	b	f
s	0.5367	0.1559	0.3074
b	0.2667	0.1963	0.5370
f	0.3699	0.2737	0.3564

Jump size distribution Based on this model, once the Markov matrix assigns the growth cone to one of the three dynamic states, the magnitude of

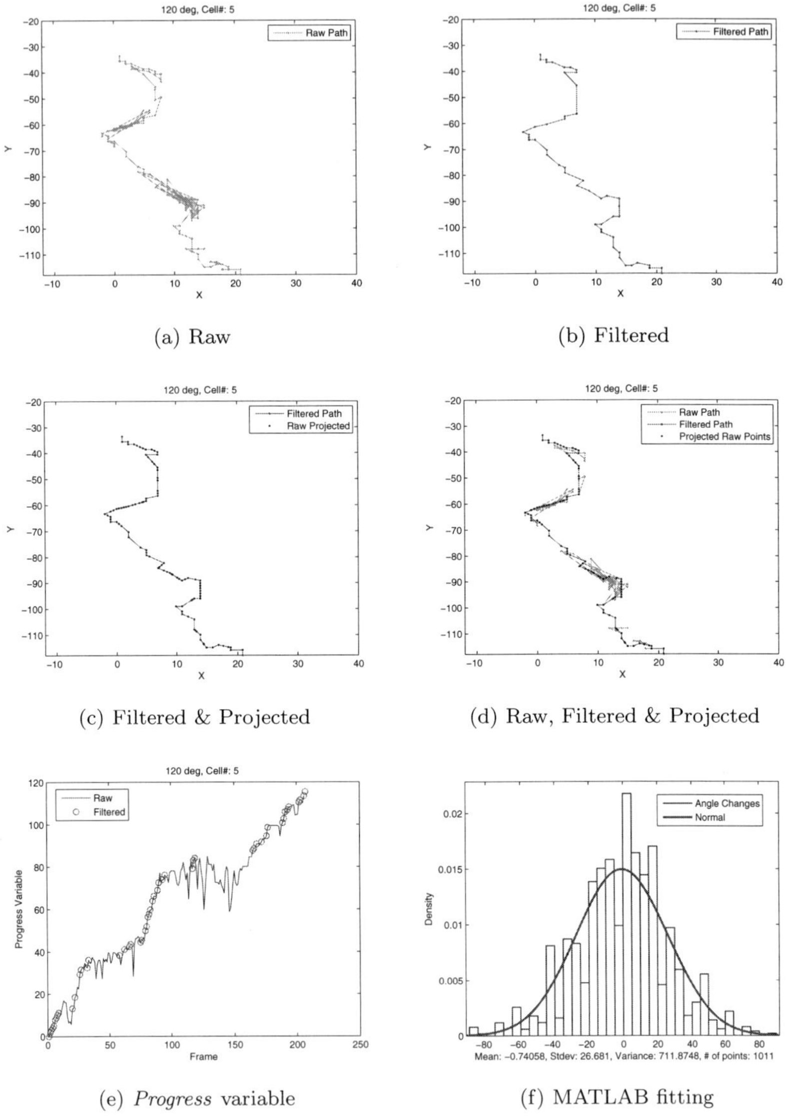

(a) Raw

(b) Filtered

(c) Filtered & Projected

(d) Raw, Filtered & Projected

(e) *Progress* variable

(f) MATLAB fitting

Fig. 2. Neuronal cell5 with 120° filtering & its $\Delta\theta$ distribution

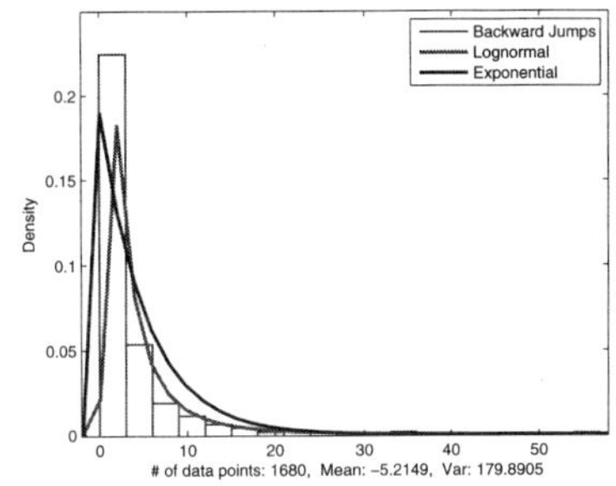

(a) LOGN(4.06,6.08) for $|\Delta S_2|$

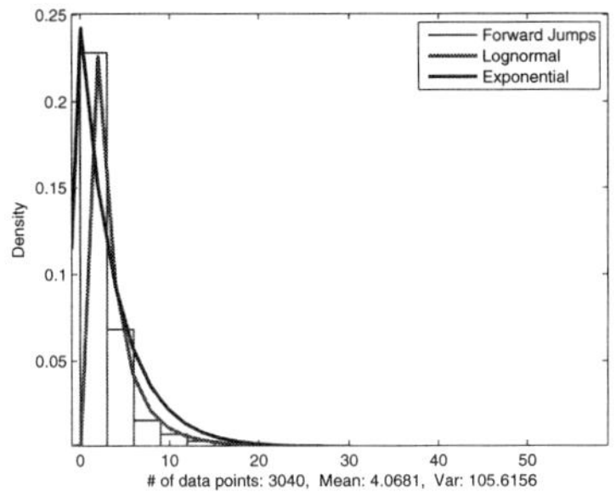

(b) LOGN(3.31, 3.53)

Fig. 3.

the growth cone motion is determined and fit with a regression to lognormal distributions separately for growth and retraction.

Conclusion

The two microscale models that we describe allow the determination of the axon dynamic parameters of direction and velocity statistics from measurements, thereby characterizing the random processes in axonal growth. Data on the dynamics of axon growth in the presence of guidance cues will determine the deterministic part of the axon growth, and whether the randomness in direction and velocity are modified by the guidance cues. Determination of direction and velocity for the growth cone depends on signaling and biocontrol processes in the filopodia and lamellipodia. We shall examine the relations between the axonal growth statistics and the statistics of the growth cone shape. This will take us one step closer to relating the genome and resulting proteome, and how they interact with the surrounding environment, with the ultimate goal a "first principles" model for neuromorphogenesis.

References

1. T. A. Lindsley, A. M. Kerlin and L. J. Rising, *Developmental Brain Research* **30**, 191 (2003).
2. A. M. Lohof, M. Quillan, Y. Dan and M. ming Poo, *The Journal of Neuroscience* **12**, 1253 (1992).
3. M. Lavigne and C. Goodman, *Science, New Series* **274**, 1123 (1996).
4. J. Sabry, T. O'Connor, L. Evans, A. Toroian-Raymond, M. Kirschner and D. Bentley, *The Journal of Cell Biology* **115**, 381 (1991).

STEADY VORTEX FLOW PAST A CYLINDER OR SPHERE

ALAN ELCRAT and KENNETH MILLER

Department of Mathematics
Wichita State University, Wichita, KS 67260
E-mail: elcrat@math.wichita.edu, miller@math.wichita.edu

BENGT FORNBERG

Department of Applied Mathematics
University of Colorado, Boulder, CO 80309
E-mail: Bengt.Fornberg@colorado.edu

The authors have found several families of axisymmetric, inviscid, stationary vortex flows which have swirling components out of the meridional plane and may be embedded in a shear flow. The flows have been obtained numerically as solutions of a partial differential equation for the stream function, the Bragg-Hawthorne equation. The vortex regions may be divided into three families which may be thought of as vortex rings with swirl, analogues of Hill's spherical vortex, and tubes of vorticity extending to infinity in both directions along the symmetry axis. Understanding the organization of these families is best done by reviewing previous results for two-dimensional vortices behind a circular cylinder and we begin by doing this. Here the vortices can be thought of as arising from the desingularization of point vortex families, e.g. the Föppl vortices. The flows that we find are either perturbations of uniform flow or uniform flow past a sphere.

1. Steady Vortices in Equilibrium with a Cylinder

Steady flows in which there is a symmetric vortex pair in equilibrium with a circular cylinder were described in Ref. 2. We first summarize those results.

For flow past a cylinder we consider the equivalent flow in a half space with a semicircular bump. We consider flows in which there is either a single region of constant vorticity or multiple regions in the case where the values of both the vorticity ($-\omega$) and the stream function on the boundary of the vortex regions (α) are the same for all regions. The flow is uniform at infinity. The stream function ψ is the solution of the nonlinear partial differential equation

$$\Delta\psi = \omega f(\psi - \alpha) \tag{1}$$

where $f = 1 - H$, and H is the Heaviside function. This corresponds to a constant vorticity ω in the region where $\psi < \alpha$, and irrotational flow elsewhere. We take $\psi = 0$ on the boundary of the flow region.

The value α of the stream function on the boundary of the vortex region is a parameter. Three types of vortex regions occur depending on the sign of α. For $\alpha = 0$ the vortex region is "attached" to the boundary, for $\alpha < 0$ the vortex region is "isolated" away from the boundary, and for $\alpha > 0$ there is a "strip" of vorticity extending to infinity along the streamline of symmetry. As α approaches $-\infty$ the vortex approaches a point vortex. The various solutions described in Ref. 2 are summarized in Fig. 1. As shown there are three families of stationary point vortex locations, each parametrized by circulation κ. Fixing κ, for each stationary point vortex location, there is an associated one parameter family of isolated vortices parametrized by α. In most cases (those point vortices indicated with bold dots in Fig. 1), α can be increased from $-\infty$ to 0, which corresponds to an attached vortex. For each attached vortex there is a sequence of strip vortices, with ω fixed, with α increasing to some maximum positive value depending on ω. The various families come together at the maximal value of α.

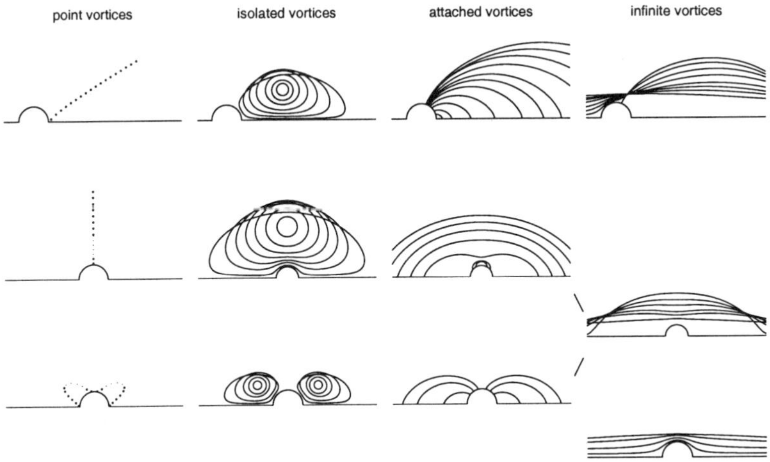

Fig. 1. One-parameter families of point vortices and attached vortices and related two-parameter families of isolated and infinite vortices. The third family of infinite vortices are perturbations of potential flow.

To compute solutions for given α and $A =$ the area of the vortex, we iteratively solved the linear Poisson equation

$$\Delta\psi_{n+1} = \omega_n f(\psi_n - \alpha), \tag{2}$$

with ω_n chosen so that the area of the vortex equaled A. (This was accomplished in an inner iteration with ω raised or lowered to achieve the correct area. The areas were approximated by using piecewise linear interpolation between grid points.) The process was repeated until the set of grid points where $\psi_{n+1} < \alpha$ was identical with the set of grid points where $\psi_n < \alpha$. To solve Eq. (2) we first transform the region exterior to the unit circle in the upper half z plane to a semi-infinite strip via the mapping $w = i\ln(z)$. The Poisson equation in the transformed domain was solved with the standard five point stencil using a uniform mesh in the w-plane on a truncation of the semi-infinite strip. On the top of the truncated strip we applied a Robin-type numerical boundary condition–a linear relation between Dirichlet and Neumann data which is exact for decaying solutions of the discrete Laplace equation on the strip. A fast Poisson solver was used at each iteration step.

2. Steady Axisymmetric Vortices in Three Dimensions

For axisymmetric flows the methods used and results are similar to the case of flow past a cylinder, with some important differences. First the differential equation for the Stokes stream function is now

$$L\psi = \omega r^2 f(\psi - \alpha) \tag{3}$$

where again $f = 1 - H$, and L is the operator

$$L\psi = \frac{\partial^2\psi}{\partial z^2} + r\frac{\partial}{\partial r}(\frac{1}{r}\frac{\partial\psi}{\partial r})$$

in cylindrical coordinates.

As shown in Ref. 3, for axisymmetric flow there are again three types of solutions: vortex rings ($\alpha < 0$), bounded vortices attached to the sphere ($\alpha = 0$) and vortex "tubes" extending to infinity along the axis of symmetry ($\alpha > 0$). One important difference between flow past a sphere and flow past a cylinder is that there is no analogue of the stationary point vortex in axisymmetric flow: infinitesimal vortex rings propagate with infinite speed.[5] Starting from an attached vortex and letting α decrease from 0, while keeping circulation κ fixed, gives a corresponding family of vortex rings. As $-\alpha$ gets larger the centers of the thin vortex rings slowly moves further away from the axis of symmetry, slowly going to infinity. This behavior is consistent with Kelvin's formula for thin vortex rings.[5]

In recent work we have extended these results to include axisymmetric flows with shear and swirl. Swirl can be included by replacing Eq. (3) with the equation

$$L\psi = r^2 f(\psi) - C(\psi)\frac{dC}{d\psi}, \tag{4}$$

where $C = rv_\theta = C(\psi)$ is the swirl.[1] The form of the profile function C used in the following computations is

$$C(\psi) = \lambda(\alpha - \psi)_+ \tag{5}$$

where λ is a swirl parameter. We can include shear in the flow by taking

$$f(\psi) = \sigma + \omega H(\alpha - \psi) \tag{6}$$

where σ is a shear parameter.

Let ψ_0 denote the stream function, vanishing on the axis of symmetry, for the background flow. Then ψ_0 satisfies

$$\begin{aligned} L\psi_0 &= \sigma r^2, && \text{for } r^2 + z^2 > a^2 \\ \psi_0 &= 0, && \text{if } r^2 + z^2 = a^2 \text{ or } r = 0 \\ \psi_0 &\sim \frac{r^2}{2} + \frac{\sigma r^4}{8}, && \text{as } r^2 + z^2 \to \infty, \end{aligned} \tag{7}$$

where a is the radius of the spherical obstacle. Setting $u = \psi - \psi_0$, it follows from Eq. (4) and Eq. (5) that we are looking for solutions u to

$$Lu = \begin{cases} r^2\omega + \lambda^2(\alpha - (u + \psi_0)), & \psi < \alpha \\ 0, & \psi > \alpha. \end{cases} \tag{8}$$

Basically, our iterative scheme for solving this equation is

$$(L + \lambda^2 I_n)u_{n+1} = (r^2\omega + \lambda^2(\alpha - \psi_0))I_n \tag{9}$$

where I_n is the characteristic function of the set $D_n = \{\psi_n < \alpha\}$. We wish to solve Eq. (9) using finite differences and then iterate until the set of grid points where $u_{n+1} < \alpha$ is identical with the set of grid points where $u_n < \alpha$. Rather than fixing area A as in the two-dimensional problem, for this problem we require that the first moment M of D_n with respect to the axis of symmetry have a prescribed value. The value of one of the three parameters ω, λ or α is then adjusted in an inner iteration so as to achieve this. Thus there are four free parameters to the general problem, the shear parameter σ, M, and two of the three parameters ω, λ and α, the remaining one of these parameters being determined as part of the solution.

When there is no obstacle, by a similarity transformation one of the non-zero parameters may be set to 1, so instead of four independent parameters there are three. Without the obstacle we have found at most one solution for each set of parameters.

We again transform the region via the mapping $\xi + i\eta = i\ln(z + ir)$ to the strip $-\pi < \xi < 0,\ 0 < \eta < \infty$. Equation (3) then transforms to

$$\tilde{L}\psi = (\omega e^{4\eta}\sin^2\xi)f(\psi - \alpha) \tag{10}$$

where $\tilde{L}$ is the differential operator

$$\tilde{L}\psi = \psi_{\xi\xi} + \psi_{\eta\eta} - (\cot\xi)\psi_\xi - \psi_\eta.$$

For swirling flow, the differential equation (9) for $u = u_{n+1}$ transforms to

$$\tilde{L}u + e^{2\eta}\lambda^2 I_n u = (\omega e^{4\eta}\sin^2\xi + e^{2\eta}\lambda^2(\alpha - \psi_0))I_n. \tag{11}$$

The boundary condition on three sides of the computational strip is $u = 0$. As explained in Ref. 3, a reasonable numerical boundary condition to impose on the upper boundary $\eta = \log R$ is the Robin condition, $\frac{\partial u}{\partial \eta} + 2u = 0$. We discretize Eq. (11) using the standard stencil for a uniform N_1 by N_2 grid on the computational rectangular. We use the sparse matrix functionality of MATLAB to solve the resulting N_1N_2 by N_1N_2 linear system at each iteration step.

3. Non-swirling flows

For flows without swirl or shear we found[3] four families of attached vortices: (a) vortex wakes behind the sphere, (b) bands of vorticity surrounding the circumference of the sphere perpendicular to the axis, (c) spherically annular vortices surrounding the sphere, and (d) symmetric vortices in front of and behind the sphere. When there is shear the same result holds, except the outer boundary of an annular vortex is no longer spherical when there is shear. Figure 2 shows examples from the four families. Associated with each attached vortex is a family of vortex rings parametrized by $\alpha < 0$. The general effect of increasing shear is to flatten and stretch out large cross-section vortices. However, small cross-section vortices are almost circular with or without shear.

4. Vortex rings with swirl

We have found families of vortices with swirl connecting each non-swirling vortex flow to a Beltrami flow (a flow in which vorticity and velocity are

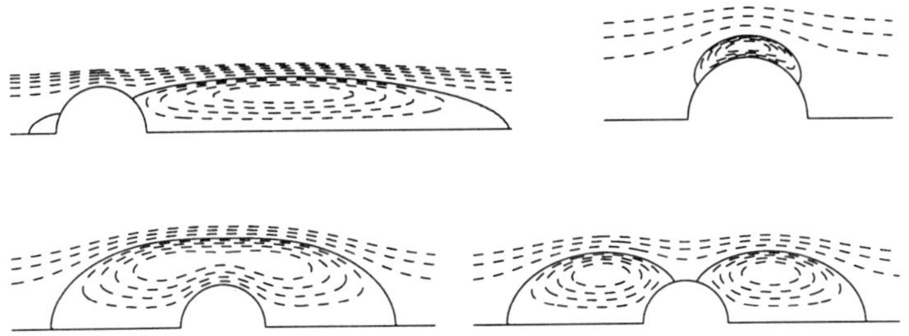

Fig. 2. Examples of the four types of attached vortices ($\alpha = 0$) without swirl($\lambda = 0$). Each is embedded in shear flow. The radius of the sphere is $a = 1$ in each case. Each subplot shows streamlines in the meridional plane. Streamlines are dashed except for the vortex boundary which is solid. Flow parameters for the examples shown are: (a), trailing vortex, $\sigma = 14$, $M = 4$; (b), vortex band, $\sigma = 0.5$, $M = 0.66$; (c), surrounding annular vortex, $\sigma = 0.5$, $M = 12$; (d), symmetric pair of vortices behind and in front of the sphere, $\sigma = 0.5$, M(total)= 7.5.

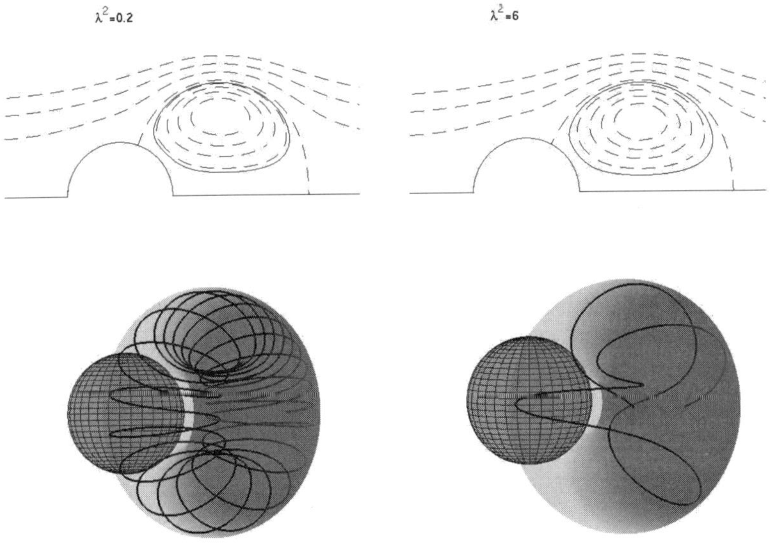

Fig. 3. Both flows shown are for $\alpha = -0.15$, $\sigma = 0$ and $M = 4$, but with $\lambda^2 = 0.2$ on the left and $\lambda^2 = 6$ on the right. (The Beltrami flow for this family has $\lambda^2 = 6.25$). The meridional plane cross sections of the flows are very similar, although not identical. In the three dimensional plots the vortex core is shaded and part of a single interior streamline is shown. The streamline is shown for 20 revolutions around the toroidal surface for the ring on the left, but for only 4 revolutions for the ring on the right, showing the increase in the variation in θ per revolution for the flow with larger $|\lambda|$.

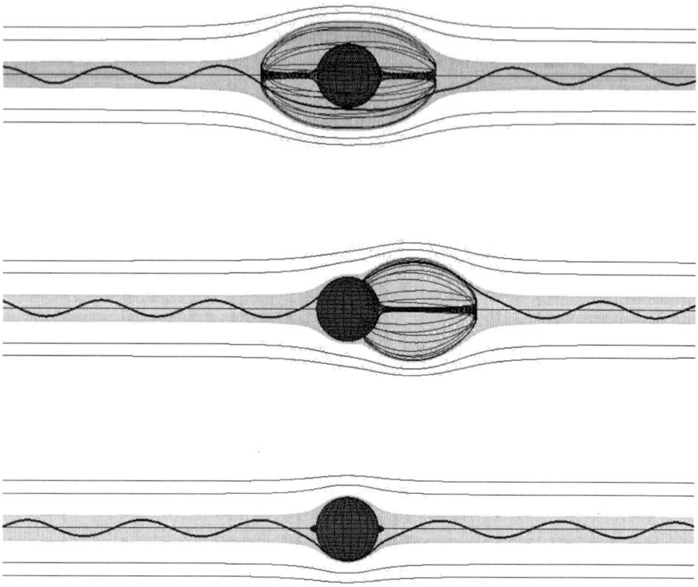

Fig. 4. Three vortex tubes for $\omega = 2$, $\lambda^2 = 1.5$, $\sigma = 0$ and $\alpha = .075$. The first tube is obtained by continuation from a symmetric, annular attached vortex with the same ω, σ and λ; the second by continuation from a trailing vortex; the third row is a perturbation of the background irrotational flow.

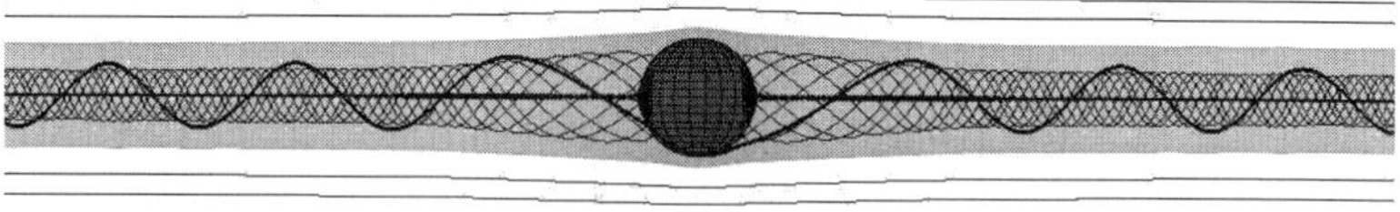

Fig. 5. Near the maximum α solution for the families in Fig. 4. The value of α is slightly less than 0.2. Two streamlines are shown in the lower plot. The helical streamline $\psi = c_1 > 0$, is printed somewhat thicker. The streamline for $\psi = c_2 < 0$ (shown for 9 loops) follows the helical streamline closely until a point well beyond the margins of the plot where it is pulled into the inner part of the toroidal surface $\psi = c_2$. That inner part of the surface appears in the plot as a thickening of the axis of symmetry.

parallel). In each case, the swirl parameter cannot be increased beyond the value which corresponds to Beltrami flow. For vortex rings, as the swirl parameter increases, the cross sections in the meridional plane only vary slightly. However, the flow patterns inside the stream surface bounding the vortex varies considerably, since the pitch of the angle with which streamlines go around a toroidal surface $\psi = c$ increases with $|\lambda|$. One example is shown in Fig. 3.

If there is no shear and $\alpha = 0$, then there is a family of exact solutions with a sphere of radius b as the outer vortex boundary. When there is no spherical obstacle such solutions were given by Moffatt.[4]

5. Vortex tubes

Solutions of Eq. (8) obtained with $\alpha > 0$ have vortex support extending to infinity along the axis of symmetry. We refer to them as vortex tubes. If $\lambda \neq 0$, then for any streamline value c, $0 < c < \alpha$, the corresponding streamlines will have a helical shape, distorted by the sphere and the recirculating core of the vortex. One such streamline is plotted in each three dimensional plot of a vortex tube. For $c < 0$, the streamlines stay in a bounded wake region; for $c > \alpha$ each streamline is non-helical. Three vortex tubes, all with the same values of ω, λ, σ and α, are shown in Fig. 4. The first two are perturbations of a symmetric attached vortex and a trailing attached vortex, respectively. The third is obtained by perturbation from the background irrotational or uniform shear flow. There is a value of α, the maximum value for which solutions exist, at which all three families come together. Figure 5 shows a solution for α near the maximal value. The recirculation region extends beyond the margins of the plot.

References

1. Batchelor, G. K., 1967 *An Introduction to Fluid Dynamics*, Cambridge Univ Press.
2. Elcrat, A., Fornberg, B., Horn, M., & Miller, K. 2000. Some steady vortex flows past a circular cylinder, *J. Fluid Mech.* 409: 13–27
3. Elcrat, A., Fornberg, B., & Miller, K. 2001. Some steady axisymmetric vortex flows past a sphere, *J. Fluid Mech.* 433: 315–328
4. Moffatt, H. K. 1969. The degree of knottedness of tangled vortex lines, *J. Fluid Mech.* 35: 117–129.
5. Saffman, P. 1992. Vortex Dynamics, Cambridge Press, Cambridge.

ON TWO FAST ALGORITHMS FOR ESTIMATING THE MIXING DISTRIBUTION IN MIXTURE MODELS

JAYANTA K. GHOSH*†‡ and RYAN MARTIN†

Department of Statistics, Purdue University†
West Lafayette, IN 47907, USA
**E-mail: ghosh@stat.purdue.edu*

Division of Theoretical Statistics and Mathematics‡
Indian Statistical Institute
Kolkata, India 700108

High dimensional mixture models are very important in bioinformatics. One major area of application is to DNA microarray data. The most difficult part of inference is the estimation of the mixing distribution. Our goal is to estimate the prior/mixing density f based on data observed from the marginal/mixture density $\pi_f(x) = \int p(x|\theta)f(\theta)\,d\mu(\theta)$. We discuss properties of two relatively new and very fast algorithms for a NP Empirical Bayes estimate by M. Newton. These estimates are compared in simulations to the NP Bayes estimate. We provide motivation through stochastic approximation and prove convergence.

Keywords: mixture models, empirical Bayes, stochastic approximation

1. Introduction

Mixture distributions have played a very important role in modeling data that reflect population heterogeneity or involve latent variables. These models have been recently used in many research areas, including genetics and bioinformatics.[1] Computational tools, such as the EM and MCMC algorithms, are now available for fitting mixture models, but estimation of the mixing distribution remains a difficult problem.

The problem can be concisely stated as follows. Suppose the observed data $x_1, \ldots, x_n$ arise from the hierarchical model

$$\theta^i \overset{\text{iid}}{\sim} f \quad \text{and} \quad x_i|\theta^i \overset{\text{ind}}{\sim} p(\cdot|\theta^i), \quad i = 1, \ldots, n \tag{1}$$

where $p(x|\theta)$ is a known sampling density on $\mathcal{X}$ with respect to a σ-finite measure ν, and f is the *unknown* mixing/prior density for θ on Θ with respect to a σ-finite measure μ. Here the θ^i's are not observed so the model in Eq. (1) is equivalent to

$$x_1, \ldots, x_n \overset{\text{iid}}{\sim} \pi_f(x) = \int_\Theta p(x|\theta) f(\theta)\, d\mu(\theta). \tag{2}$$

The goal is to estimate f based on the indirect observations $x_1, \ldots, x_n$.

Typically, $\mathcal{X}$ and Θ are Euclidean spaces and ν is either Lebesgue or counting measure. The choice of μ depends on the desired inference: for estimation of $\theta^1, \ldots, \theta^n$, μ is Lebesgue or counting measure, but for testing, μ is usually decomposed as $\mu = \mu_0 + \mu_1$ where μ_0 and μ_1 are measures dominating the null and alternative distributions, respectively.

Applications of the model in Eq. (1) include the analysis microarray data on gene expression. Here θ^i represents the expression level of the ith gene under investigation, with $\theta^i = 0$ indicating that the ith gene is not differentially expressed. The model is as in Eq. (1) where f is a prior density with respect to $\mu = \lambda_{\text{Leb}} + \delta_{\{0\}}$. Consider the multiple testing problem

$$H_{0i} : \theta^i = 0, \quad i = 1, \ldots, n.$$

The number n of genes being investigated is often in the thousands and, in such cases, classical procedures based on the individual x_i's can be considerably improved by using all the available data. Also, the results of a fully Bayesian analysis are highly sensitive to the choice of prior f. Interestingly, while the data $x_1, \ldots, x_n$ contains little information about the individual parameters, for large n there is considerable information about the prior. In such cases, the *empirical Bayes* approach[2]—where the data is used to *estimate* the prior—can be applied successfully to this problem.[3]

For estimation of the mixing/prior density f, we focus our attention on a relatively new algorithm proposed by Newton *et al.*[4,5]

Algorithm 1.1. *Fix a suitable initial estimate f_0 and a weight sequence $w_1, w_2, \ldots \in (0, 1)$. Given iid observations $x_1, \ldots, x_n$ from the mixture density $\pi_f(x)$ in Eq. (2), compute*

$$f_i(\theta) = (1 - w_i) f_{i-1}(\theta) + w_i \frac{p(x_i|\theta) f_{i-1}(\theta)}{\int_\Theta p(x_i|\theta') f_{i-1}(\theta')\, d\mu(\theta')}, \quad \theta \in \Theta \tag{3}$$

for $i = 1, \ldots, n$ and produce f_n as the final estimate.

In later sections we will discuss properties of f_n in more detail, but a few important observations are immediate from the definition.

(iii) f_n depends on the order in which the data enters the recursion.

(iiiii) Prior information concerning the support or shape of f can be incorporated quite easily through the initial guess f_0.

(iiiiiiii) Algorithm 1.1 is more general than deconvolution methods as it can be applied to any $p(x|\theta)$, not simply those of the form $p(x-\theta)$.

(iviviv) The algorithm is very fast: if f_n is computed on a grid of m values of θ, then the computational complexity is mn.

The fact that f_n depends on the order of the data is a bit troublesome. In specific cases,[6] an "appropriate" permutation of the data may suggest itself. In general, however, a *permutation invariant* counterpart to f_n is desirable. Newton[5] suggested using a pointwise average $\hat{f}_n$ of the f_n's over several randomly chosen permutations. This $\hat{f}_n$ can be seen as a Monte Carlo approximation to the full average

$$\overline{f}_n(\theta) = \mathbb{E}[f_n(\theta) \mid x_{(1)}, \ldots, x_{(n)}] = \frac{1}{n!} \sum_{s \in \mathcal{S}_n} f_{n,s}(\theta), \tag{4}$$

where $\mathcal{S}_n$ is the permutation group on $\{1, 2, \ldots, n\}$ and $f_{n,s}$ is the estimate f_n with the data arranged as $x_{s(1)}, \ldots, x_{s(n)}$.

The rest of the paper is organized as follows. Section 2 contains two distinct motivations for the recursive algorithm. In Sec. 3 we review the convergence properties of the algorithm and its representation as a stochastic approximation. In Sec. 4, simulation results are presented that demonstrate the accuracy and efficiency of f_n and $\hat{f}_n$ compared to one of its most popular competitors, the nonparametric Bayes estimates.

2. Motivation

In this section we discuss the two primary motivating factors behind the use of the recursive estimate f_n.

Approximately Bayes. The recursive estimate is not a posterior quantity (see Remark (i)), but the substantial flexibility resembles what one finds in a Bayesian framework. In fact, the original motivation[4] for Algorithm 1.1 was to give a heuristic, yet computationally efficient, approximation to the popular nonparametric Bayes estimate based on a Dirichlet process (DP) prior. While simulations presented in Refs. 7–9 and in Sec. 4 suggest that f_n is not an approximation of the DP prior Bayes estimate, it is still possible for it to be an approximation of some other Bayes estimate.

Stochastic approximation. Stochastic approximation (SA) is an algorithmic method for locating a root ξ of a function h when only noisy observations on h are available; *e.g.*, in the regression problem where, for any input x, one observes $y = h(x) + \varepsilon$, rather than $h(x)$ itself. Robbins and Monro[10] proposed the following algorithm for constructing a sequence $\{x_n\}$ that converge to ξ. Let $\{\varepsilon_n : n \geq 1\}$ be a white noise process.

Algorithm 2.1. *Choose an initial guess x_0 of ξ. At time $n \geq 1$, record the noisy observation $y_n = h(x_{n-1}) + \varepsilon_n$ of $h(x_{n-1})$ and set*

$$x_n = x_{n-1} + w_n\, y_n, \tag{5}$$

where the weight sequence $\{w_n\}$ satisfies

$$w_n > 0, \quad \textstyle\sum_n w_n = \infty, \quad \textit{and} \quad \sum_n w_n^2 < \infty. \tag{6}$$

The condition $\sum_n w_n = \infty$ on the weights is necessary for x_n to outgrow the influence of the initial guess x_0. The condition $\sum_n w_n^2 < \infty$ forces $w_n \to 0$, allowing for accumulation of information, as well as ensuring a certain rate of decay. For more details on SA, see Ref. 11.

We claim that Newton's recursive estimate f_n in Eq. (3) can be represented as a SA algorithm. Towards this, define the operator

$$H(x,\varphi)(\theta) = \frac{p(x|\theta)\varphi(\theta)}{\int p(x|\theta')\varphi(\theta')\, d\mu(\theta')} - \varphi(\theta), \tag{7}$$

where $x \in \mathcal{X}$ and φ is a density on Θ with respect to μ. Simple algebraic manipulation shows that Eq. (3) is equivalent to

$$f_n(\theta) = f_{n-1}(\theta) + w_n H(x_n, f_{n-1})(\theta), \quad n \geq 1. \tag{8}$$

Associating y_n in Algorithm 2.1 with $H(x_n, f_{n-1})$ in Eq. (7) shows that f_n is a type of SA algorithm. Letting $h(\varphi)$ be the expectation of $H(x,\varphi)$, we expect the iterates f_n to converge to a solution of the equation $h(\varphi) = 0$, where the operator h is defined as

$$h(\varphi) = \mathbb{E}_{\pi_f}[H(x,\varphi)] = \int_{\mathcal{X}} \frac{\pi_f(x)}{\pi_\varphi(x)} p(x|\theta)\varphi(\theta)\, d\nu(x) - \varphi(\theta) \tag{9}$$

It is easy to see that $\varphi = f$ is a solution to $h(\varphi) = 0$. In Sec. 3 we use a general SA theorem to prove, in the case of finite Θ, that $f_n \to f$ a.s. for any choice of initial guess f_0.

Remark 2.1. The SA approach above and, particularly, the map $h(\varphi)$ in Eq. (9), is closely related to the so-called I-projections of Ref. 12.

3. Theoretical properties

In this section, we review the convergence properties of the estimates f_n and $\overline{f}_n$. Theorem 3.1[9] is the most general result known. We sketch a proof of Theorem 3.2[8] based on ODE stability results and the SA representation of f_n in Sec. 2.

Theorem 3.1 requires the following assumptions.

A1. The weights $w_1, w_2, \ldots \in (0,1)$ satisfy the conditions in Eq. (6).
A2. f is *identifiable*; i.e. $\pi_\varphi = \pi_\psi$ ν-a.e. implies $\varphi = \psi$ μ-a.e.
A3. For each $x \in \mathcal{X}$, the map $\theta \mapsto p(x|\theta)$ is bounded and continuous.
A4. For any $\varepsilon > 0$ and any compact set $\mathcal{X}_0 \subset \mathcal{X}$, there exists a compact set $\Theta_0 \subset \Theta$ such that $\int_{\mathcal{X}_0} p(x|\theta)\, d\nu(x) < \varepsilon$ for all $\theta \notin \Theta_0$.
A5. There exists a constant $B < \infty$ such that, for every $\theta_1, \theta_2, \theta_3 \in \Theta$

$$\int_{\mathcal{X}} \left(\frac{p(x|\theta_1)}{p(x|\theta_2)} \right)^2 p(x|\theta_3)\, d\nu(x) < B.$$

Conditions A3 and A4 are readily checked for many common densities $p(x|\theta)$, such as Normal, Gamma, Poisson; A5 is satisfied in these cases if Θ is a compact interval.

Theorem 3.1. *Under* A1–A5, $f_n \to f$ *a.s. and* $\overline{f}_n \to f$ *in probability as* $n \to \infty$*; in both cases, convergence is in the weak topology.*

The proof of this theorem is based on an approximate martingale representation of the two Kullback-Leibler (KL) divergences $K(f, f_n)$ and $K(\pi_f, \pi_n)$, where $\pi_n(x) = \int p(x|\theta) f_n(\theta)\, d\mu(\theta)$. Specifically, it is shown that $K(\pi_f, \pi_n) \to 0$ a.s. and then a tightness argument is used to extend L_1 consistency of π_n to weak consistency of f_n. The claim for $\overline{f}_n$ is proved similarly—the key observation is the inequality $\mathbb{E}[K(f, \overline{f}_n)] \leq \mathbb{E}[K(f, f_n)]$.

Theorem 3.2. *Let* μ *be counting measure on* $\Theta = \{\theta_1, \ldots, \theta_d\}$ *and assume* A1–A2 *and* $p(\cdot|\theta) > 0$ ν*-a.e. for each* θ*. Then* $f_n \to f$ *a.s.*

Proof Outline. Let $\varphi = (\varphi^1, \ldots, \varphi^d)'$ be a generic vector in the probability simplex Δ, and let $f = (f^1, \ldots, f^d)' \in \Delta$ be the true mixing distribution. The proof consists of showing that f is an asymptotically stable[13] equilibrium point of the associated ODE $\dot{\varphi} = h(\varphi)$ and then appealing to the general SA theorems[11] which say that SA sequences as in Eq. (8) have the same sample path properties as solutions to $\dot{\varphi} = h(\varphi)$ a.s.

To this end, consider the Kullback-Leibler divergence

$$\ell(\varphi) = K(f, \varphi) = \textstyle\sum_{i=1}^d f^i \log(f^i/\varphi^i).$$

Assume, for simplicity, that $f \in \text{int}(\Delta)$; the result holds without this assumption. Clearly, $\ell(\varphi)$ is positive definite and continuously differentiable in a neighborhood of f. Its gradient can be written as

$$\nabla \ell(\varphi) = -(r^1, \ldots, r^d)' + r^d 1_d$$

where $r^k = f^k/\varphi^k$ and 1_d is a d-vector of unity. The time derivative $\dot{\ell}(\varphi)$ along a trajectory φ of the associated ODE is

$$\dot{\ell}(\varphi) = \nabla \ell(\varphi)' h(\varphi) = \int \frac{\pi_f(x) - \pi_\varphi(x)}{\pi_\varphi(x)} \nabla \ell(\varphi)' P_x \varphi \, d\nu(x)$$
$$= \int \frac{\pi_\varphi - \pi_f}{\pi_\varphi} \pi_f \, d\nu = 1 - \int \frac{\pi_f}{\pi_\varphi} \pi_f \, d\nu$$

where $P_x = \text{diag}\{p(x|\theta_1), \ldots, p(x|\theta_d)\}$ is the $d \times d$ diagonal matrix of sampling density values. Applying Jensen's inequality to the last expression, along with A2, shows that $\dot{\ell}(\varphi) \leq 0$ with equality iff $\varphi = f$. This proves that $\ell(\varphi)$ is a Lyapunov function and, since $\dot{\ell}(\varphi)$ vanishes only at f, any trajectory of $\varphi(t)$ satisfying $\dot{\varphi} = h(\varphi)$ converges to f as $t \to \infty$. From the connection between SA and ODEs,[11] it follows that $f_n \to f$ a.s. □

4. Simulations

In this section we compare the performance of the recursive estimate (RE) and the recursive estimate averaged over permutations (PARE) with that of the nonparametric Bayes estimate (NPBE) based on a Dirichlet process prior $\mathcal{D}(1, f_0)$ with base measure f_0 and precision 1. First we define our simulation parameters:

- 100 samples of size $n = 200$ are taken from the model.
- For PARE, 100 random permutations of the data are selected.
- For RE and PARE, the weights are $w_i = (i+1)^{-1}$ for $i = 1, \ldots, n$.
- For RE, PARE and NPBE, f_0 is a Unif(Θ) density.

Consider the mixture model where, for $i = 1, \ldots, n$,

$$\theta_i \overset{\text{iid}}{\sim} \tfrac{1}{3}\,\text{Beta}(3, 30) + \tfrac{2}{3}\,\text{Beta}(4, 4), \quad \text{and} \quad x_i|\theta_i \overset{\text{ind}}{\sim} N(\theta_i, \sigma^2).$$

One can easily check that conditions A2–A5 hold. Here we choose $\sigma = 0.1$ but our conclusions are fairly robust to this choice. Each cell of Fig. 1 shows the true density as well as the 100 estimates. We see that the NPBE of f can be quite poor (top row) but the corresponding estimate of π_f performs much better (bottom row). On the other hand, f_n and $\hat{f}_n$ sit relatively close to f across the 100 samples. As expected from the inequality

$\mathbb{E}[K(f, \overline{f}_n)] \leq \mathbb{E}[K(f, f_n)]$ mentioned in Sec. 3, we see less variablility in the PARE than in the RE. This observation is even more striking when one looks at the corresponding marginals. In Fig. 2, we see that RE and NPBE perform comparably, with PARE being considerably more accurate than the others, in a KL sense. We should also point out that PARE is, on average, about 10 times faster to compute than NPBE.

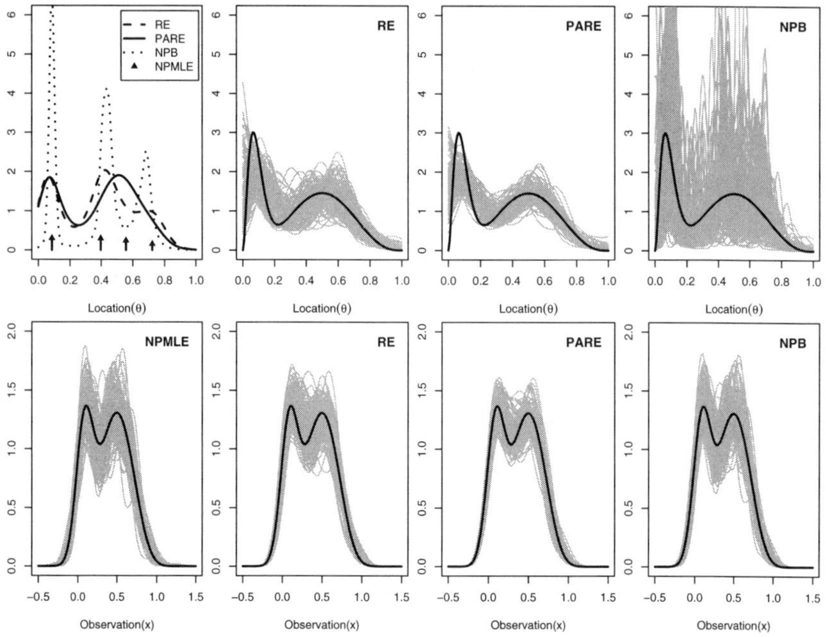

Fig. 1. Plots of the true mixing density f (black, top row), the true mixture density π_f (black, bottom row) and 100 corresponding estimates (gray).

References

1. D. Allison, G. Gadbury, M. Heo, J. Fernández, C. Lee, T. Prolla and R. Weindruch, *Comput. Statist. Data Anal.* **39**, 1 (2002).
2. H. Robbins, *Ann. Math. Statist.* **35**, 1 (1964).
3. B. Efron, R. Tibshirani, J. Storey and V. Tusher, *J. Amer. Statist. Assoc.* **96**, 1151 (2001).
4. M. Newton, F. Quintana and Y. Zhang, Nonparametric Bayes methods using predictive updating, in *Practical Nonparametric and Semiparametric Bayesian Statistics*, eds. D. Dey, P. Muller and D. Sinha (Springer, New York, 1998)

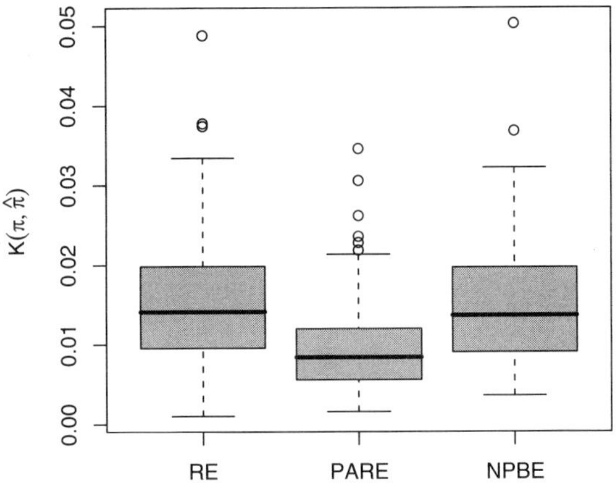

Fig. 2. Summary of the KL divergences $K(\pi_f, \hat{\pi})$ over the 100 samples, where $\hat{\pi}$ is the RE, PARE or NPBE of π_f.

5. M. Newton, *Sankhya* **A 64**, 306 (2002).
6. M. Bogdan, J. Ghosh and S. Tokdar, A comparison of the Benjamini-Hochberg prodecure with some Bayesian rules for multiple testing, in *Festschrift for P.K. Sen. IMS Lecture Notes–Monograph Series*, ed. M. Silvapulle (IMS, Beachwood, OH, 2007) To appear.
7. J. Ghosh and S. Tokdar, Convergence and consistency of Newton's algorithm for estimating a mixing distribution, in *The Frontiers of Statistics*, eds. J. Fan and H. Koul (Imperial College Press, London, 2006) pp. 429–443.
8. R. Martin and J. Ghosh, Stochastic approximation and Newton's estimate of a mixing distribution, Submitted, (2007).
9. S. Tokdar, R. Martin and J. Ghosh, Consistency of a recursive estimate of mixing distributions, Submitted, (2008).
10. H. Robbins and S. Monro, *Ann. Math. Statist.* **22**, 400 (1951).
11. H. Kushner and G. Yin, *Stochastic Approximation and Recursive Algorithms and Applications*, 2nd edn. (Springer, New York, 2003).
12. N. Shyamalkumar, *Cyclic I_0 projections and its applications in statistics*, Tech. Rep. 96-24, Purdue University (Department of Statistics, West Lafayette, IN, 1996).
13. J. LaSalle and S. Lefschetz, *Stability by Liapunov's Direct Method with Applications* (Academic Press, New York, 1961).

DIRECT REGRESSION MODELS FOR SURVIVAL PARAMETERS BASED ON PSEUDO-VALUES*

JOHN P. KLEIN[‡] and GISELA TUNES-DA-SILVA[†]

Division of Biostatistics, Department of Population Health, Medical College of Wisconsin
8701 Watertown Plank Road
Milwaukee, WI 53226
[‡]*E-mail: klein@mcw.edu*

We investigated the use of pseudo-values from a jackknife statistic constructed from a simple summary statistic as a way of developing direct regression models of survival parameters. These pseudo-values, based on the difference between the complete sample and leave-one-out estimator, are used in a generalized estimating equation to obtain estimates of model parameters. The approach can be applied to direct regression modeling of the survival function over time, the cumulative incidence function for competing risk data, the restricted mean survival time, the mean quality of life, and the probabilities in a multistate model.

1. Introduction

Semi-parametric regression models for right censored data typically involve modeling hazard rates or intensity functions. Of theses models the most common is the model of Cox (Ref. 1) which models the log of the survival function as a linear function of covariates. This model is also used to model the intensity functions in a multistate model (Ref. 2). The model can also be applied to modeling crude hazard rates in a competing risks framework (cf Ref. 3). Other semi parametric approaches model covariate effects on the intensity function as an additive model (cf Ref. 4 or Ref. 5) or a mixed model (cf. Ref. 6). For a survey of these and other models see Ref. 7.

Most regression models for survival data focus on a model for the entire intensity function. When the outcome is death this provides a model for

*This research was partially supported by a grant (R01 CA54706-13) from the National Cancer Institute and by a grant (07/02823-3) from FAPESP, Brazil

[†]Also at Department of Statistics, University of São Paulo, São Paulo, Brazil

the entire survival curve. A regression model for the survival function at a single time point in these models requires assuming a model for the entire curve. When the outcome is from a competing risks experiment or when it is a state or transition probability in a multistate model the quantity of interest is typically a complex non-linear function of several transition intensities. This makes it difficult to determine what the effect of a given covariate is on a particular probability.

In this note we review a general approach to censored data regression first proposed by Andersen and Klein (Ref. 8). This approach allows direct modeling of many common survival parameters. It is based on pseudo-observations constructed from an approximately unbiased univariate estimator of the parameter of interest. These pseudo-observations are then used in a generalized estimating equation to estimate the regression models. We then illustrate the approach in a number of common survival applications.

2. Pseudo Observation Regression

Let $\{X_i, i = 1, \ldots, n\}$ be independent and identically distributed random variables with distribution on some sample space. The X_i's could be random variables, a vector $X_i = \{X_{i1}, \ldots, X_{iK}\}$ or a stochastic process $X_i = \{X_i(t), t > 0\}$. We assume that we are interested in a regression model for the expectation, θ, of some function $f(X)$. We allow for the possibility that $f(\cdot)$ may be a multivariate function. Note that we have

$$\theta = \mathrm{E}\left[f(X)\right] = \int f(x)P(dx).$$

Let $\hat{\theta} = \hat{\theta}(X)$ be an unbiased estimator of θ. That is

$$\mathrm{E}_P[\hat{\theta}] = \int \hat{\theta}(x)P(dx) = \theta.$$

Now suppose we have a set of covariates $Z_1, \ldots, Z_n$ which are an independent and identically distributed sample from a distribution Q on the same sample space. Let R be the joint distribution of (X, Z). Then we have

$$\theta = \int f(x)P(dx) = \int\int f(x)R(dx, dz) = \int \mathrm{E}\left[f(X)|Z\right]Q(dz).$$

This implies that $\hat{\theta}$ is also unbiased for θ with respect to R and θ is the Q mixture of the conditional expectation of $f(X_i)$ with respect to Z_i. We now define the random variables

$$\theta_i(Z_i) = \mathrm{E}\left[f(X_i)|Z_i\right], i = 1, \ldots, n \quad (1)$$

and their average,

$$\tilde{\theta}(Z) = \left(\frac{1}{n}\right) \sum_{i=1}^{n} \theta_i(Z_i).$$

Then we have

$$\mathrm{E}_R\left[\tilde{\theta}(Z)\right] = \mathrm{E}_Q\left[\tilde{\theta}(Z)\right] = \frac{1}{n}\sum_{i=1}^{n} \mathrm{E}_Q\left[\theta_i(Z_i)\right] = \theta \tag{2}$$

so $\tilde{\theta}(Z)$ is unbiased for θ with respect to R. Define now the leave-one-out estimator, $\tilde{\theta}_{-i}(Z)$, based on all $(n-1)$ observations with $i \neq j$, by

$$\tilde{\theta}_{-i}(Z) = \frac{1}{n-1} \sum_{i \neq j} \theta_j(Z_j).$$

We can write $\theta_i(Z_i)$ as

$$\theta_i(Z_i) = \tilde{\theta}(Z) + (n-1)\left[\tilde{\theta}(Z) - \tilde{\theta}_{-i}(Z)\right] = n\tilde{\theta}(Z) + (n-1)\tilde{\theta}_{-i}(Z).$$

By Eq. (1) and Eq. (2) we can replace $\tilde{\theta}(Z)$ in Eq. (2) by

$$\hat{\theta}_i = \hat{\theta}(X) + (n-1)\left[\hat{\theta}(X) - \hat{\theta}_{-i}(X)\right] = n\hat{\theta}(X) + (n-1)\hat{\theta}_{-i}(X), \tag{3}$$

where $\hat{\theta}_{-i}(X)$ is the leave-one-out estimator based on $\hat{\theta}(X)$. Note that

$$\mathrm{E}_R\left[\theta_i(Z_i)\right] = \mathrm{E}_R\left[\hat{\theta}_i\right].$$

The $\hat{\theta}_i$'s given by Eq. (3) are the pseudo-observations from a jackknife procedure (see Ref. 9 or Ref. 10). As shown in Ref. 8 these pseudo-observations can be used in a generalized linear model to model the effects of covariates on outcome. Let $g(\cdot)$ be a link function. We assume a generalized linear model with

$$g(\theta_i) = \beta^T Z_i, i = 1, \ldots, n.$$

Here we allow for the possibility that θ_i be a multi-dimensional and for β to be a vector valued function. Define the inverse link by

$$\theta_i = g^{-1}(\beta^T Z_i) = \mu(\beta^T Z_i).$$

We shall use the generalized estimating equation approach (GEE) of Ref. 11 to estimate β. The estimating equations to be solved are

$$U(\beta) = \sum_i \left(\frac{\partial \mu_i}{\partial \beta}\right) V_i^{-1}(\beta)\left(\hat{\theta}_i - \theta_i\right) = \sum_i U_i(\beta) = 0, \tag{4}$$

where V_i is a working covariance matrix.

Let $\hat{\beta}$ be the solution to Eq. (4). Using results from Ref. 11, under standard regularity conditions, one can show that $n^{1/2}(\hat{\beta}-\beta)$ is asymptotically normal with mean zero and a covariance that can be estimated consistently by a "sandwich" estimator given by

$$\hat{\Sigma} = I(\hat{\beta})^{-1}\text{var}\left(U(\hat{\beta})\right) I(\hat{\beta})^{-1},$$

where

$$I(\beta) = \sum_i \left(\frac{\partial \mu_i}{\partial \beta}\right)^T V_i(\beta)^{-1} \left(\frac{\partial \mu_i}{\partial \beta}\right)$$

and

$$\text{var}\left(U(\hat{\beta})\right) = \sum_i U_i(\beta)U_i(\beta)^T.$$

Alternative variance estimator can be constructed using the bootstrap.

The above theory assumes that an unbiased estimator of θ is available. With right censored and/or left censored data an approximately unbiased estimator is available which can be used. Monte Carlo results suggest that that regression models based on these estimators have good properties. In the following section we review how this approach can be used to model the survival function, the mean survival time, the cumulative incidence function, state occupation probabilities in multistate models and mean quality of life.

3. Direct Regression for Survival Function

Direct regression modeling based on the pseudo-observation approach with right censored data can be based on the Kaplan-Meier estimator. This estimator is defined by

$$\hat{S}(t) = \prod_{t_i \leq t} \left(1 - \frac{d_i}{Y_i}\right),$$

where the t_i's are the ordered event times, d_i the number of events and Y_i the number at risk at time t_i. To construct a direct regression model for the survival function we start with a grid of time points, $\tau_1, \tau_2, \ldots, \tau_k$. Monte Carlo studies (cf. Ref. 8) suggest that 5-10 time points roughly equally spaced on the event scale is sufficient. We compute the pseudo-observations $\hat{\theta}_{ih}$, $i = 1, \ldots, n$, $h = 1, \ldots, k$ at each time point. If there are no censored observations prior to time τ_m then the pseudo-observations for $h = 1, \ldots, m$ reduce to the indicator that the ith person was alive or dead at τ_i. For the working covariance matrix we typically use the identity of working independence matrix (See Ref. 12). Possible link functions are the complimentary log-log link

$$g(S(t)) = \ln[-\ln(S(t))]$$

or the logit link,

$$g(S(t)) = \ln\left[S(t)/(1 - S(t))\right].$$

The complimentary log-log link reduces to the Cox proportional hazards model when the τ's are taken to be the observed event times.

The pseudo-observation approach when applied to the survival function is perhaps most useful when there is a single time point. As noted in Ref. 13 and Ref. 14 this approach is particularly useful in comparing survival curves that may cross at some point in time. When we use the logit link and there is no censoring prior to τ this approach is equivalent to a logistic regression with a robust variance estimator.

The approach requires that the pseudo-observations be computed only once, not for each model. SAS and R functions to compute the pseudo-observations from the Kaplan-Meier estimator are available on our website at www.biostat.mcw.edu. These macros are discussed and illustrated in Ref. 15.

To illustrate the approach we consider the 21 patients treated with 6 MP reported in Ref. 16. Here we examine the effect of remission status at the time of treatment on the time to relapse or death. Table 1 shows the data and the pseudo-values obtained looking at inference for 12-month survival probability. Here we see that the pseudo-observations for censored observations prior to 12 months are up weighted while uncensored observations beyond the first censoring time are given values slightly bigger than one. Using these pseudo-observations and the complementary log link gives a estimated relative risk of death or relapse of 6.92 (95% Confidence interval: 1.09, 43.92) when comparing the complete to the partial remission cases.

4. Direct Regression Modeling for Mean Survival

A second application of the pseudo-observation approach is to the mean survival (Ref. 17). With no censoring the mean can be estimated in an unbiased way by the sample mean which is equal to the area under the empirical survival function. With censored data the Kaplan-Meier curve may not be well defined beyond the largest observation. Thus we look at a

Complete Remission Cases			Partial Remission Cases		
time†	Pseudo Observations		time†	Pseudo Observations	
	S(12)	Mean‡		S(12)	Mean‡
6	0	6	6	0	6
6+	0.88	23.55	6	0	6
7	-0.06	5.96	10	-0.13	8.05
9+	0.94	24.65	20+	1.01	28.83
10+	1.01	25.84	36+	1.01	33.16
11+	1.01	25.84			
13	1.01	8.84			
16	1.01	12.88			
17+	1.01	28.83			
19+	1.01	28.83			
22	1.01	17.01			
23	1.01	19.03			
25+	1.01	33.16			
32+	1.01	33.16			
34+	1.01	33.16			

† + indicates Censored Observation

‡ Mean Restricted to 30 months

regression model for the restricted mean, μ_τ, defined by

$$\mu_\tau = \int_0^\tau S(u)du.$$

Pseudo observations are computed by using the area under the Kaplan-Meier estimator between 0 and τ. SAS and R macros to perform these calculations are found in Ref. 15. Note when there is no censoring and τ is set to be very large the pseudo-observations are the same as the original data and we are performing linear regression.

We again illustrate on the 6-MP data described above. Table 1 includes the pseudo-observations for the mean restricted to less than 30 months. Here the pseudo-observations for censored observations are adjusted up while those for events (after the first censored value) are adjusted down. Fitting a generalized estimating equation model to this data, with an identity link function, yields an estimated effect of 6.08 months with a standard error of 5.86. This implies that complete remission patients live on average 6 months longer than partial remission patients. However, this is not a statistically significant increase in mean life (p=0.30).

5. Direct Regression Modeling for Competing Risks

A third application of the methodology is to direct regression modeling of the cumulative incidence function in competing risks data. In competing risks data we have $K \geq 2$ events with potential event times $T_1, \ldots, T_K$. We observe only the minimum of these event times, T, and an indicator, δ, which tells us which of competing risks failed first. Key quantities of interest are the crude hazard rate for cause k defined by

$$h_k(t) = \lim_{\Delta t \to 0} \frac{P(t < T < t + \Delta t, \delta = k | T \geq t)}{\Delta t}, k = 1, \ldots, K$$

and the cumulative incidence function for cause k defined by

$$C_k(t) = P(T < t, \delta = k | T \geq t) = \int_0^t h_k(u) \exp\left\{ -\sum_{j=1}^{K} \int_0^u h_j(\nu) d\nu \right\} du.$$

Most regression analysis of competing risks data is based on a proportional hazards model for $h_k(t)$ while summary curves are based on $C_k(t)$. Limited models have been proposed for direct regression modeling of $C_k(t)$ (See Ref. 18). The pseudo-observation approach can be applied to the cumulative incidence function in a similar manner as it was applied to the survival function. Here we base the pseudo-observations based on the empirical cumulative incidence function given by

$$\hat{C}_k(t) = \sum_{t_i < t} \left(\frac{d_{ik}}{Y_i} \right) \prod_{t_j \leq t_i} \left(1 - \frac{\sum_{h=1}^{K} d_{jh}}{Y_j} \right).$$

Here the t_i's are the distinct event times (regardless of cause), Y_i the number of risk, and d_{ik} the number of type k events at t_i. When there is no censoring the estimated cumulative incidence function is simply the number of failures of type k prior to t and the pseudo-value at t is simply the indicator of type k failure at time t.

An extensive study of the use of pseudo-observations for competing risks data can be found in recent papers by Klein and Andersen (Ref. 12 and Ref. 19). SAS and R macros to compute pseudo-values for the cumulative incidence function can be found in Ref. 15.

6. Regression Models for State Probabilities in Multistate Models

An important use for the pseudo-observation techniques is in direct regression modeling of state occupation probabilities of which the cumulative incidence function is a special case (See Ref. 20). Examples and theory for this type of model can be found in Ref. 8 and Ref. 19.

Figure 1 shows a typical multistate model from the Bone marrow recovery process for patients with a bone marrow transplant for Chronic Myeloid Leukemia. In this model patient right have transplant are free of disease (State 0). From this state one can die (State 1) or have a return of their disease (State 2). A new therapy called a donor leukocyte infusion or DLI gives back donor cells to the patient to induce a second remission. This second remission is state 4 from which a patient can die or have additional relapses. The process is illustrated in Fig. 1. Of interest is a regression model for, for example, the probability of being in state 4 at time t.

The pseudo-values could be obtained by either by fitting Nelson-Aalen type estimators to each of the transition rates and then constructing a Aalen-Johansen estimator form the intensity transition matrix (See Ref. 21) or by using an estimator based on the difference of Kaplan-Meier estimators (See Ref. 22). For example, here we can estimate the probability of being in state 4 as the Kaplan-Meier estimator of being in state 0, 2 or 4 obtained treating occurrences of transitions into state 1, 3 or 5 as events, $\hat{S}_{024}(t)$, minus the Kaplan-Meier estimator of being in state 0 or 2 obtained by treating occurrence of a transition into state 1 or 3 or 4 as events, $\hat{S}_{02}(t)$. Details can be found in Ref. 19.

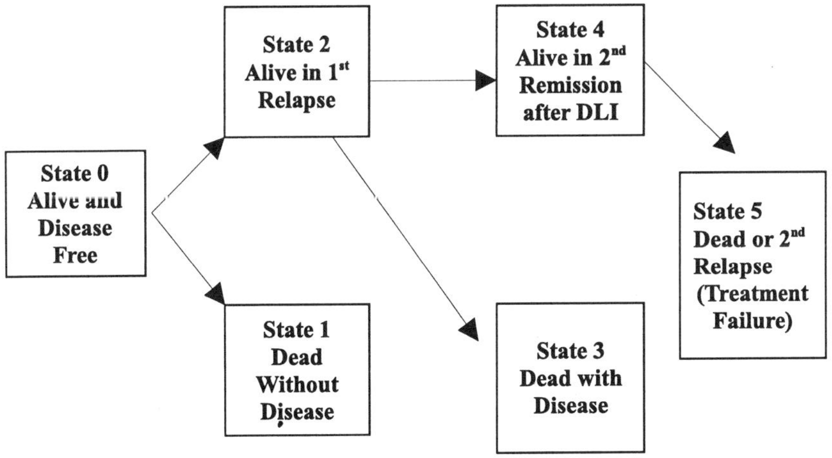

Fig. 1. Recovery process after a bone marrow transplant

7. Mean Quality-of-Life

An extension of the multistate model in the previous section is use the pseudo-observation approach to model the mean quality of life.[23] Here we assign a utility between 0 and 1 to each of the possible health states with 1 being perfect health and 0 being dead. We then estimate the expected length of time a patient is expected to be in a give health state. These estimated state occupancy times are based on a truncated mean obtained as the difference in Kaplan-Meier estimators. For example, the expected time spent in state 4 is given by

$$Y_4 = \int_0^{\mathcal{T}} [S_{024}(t) - S_{02}(t)]\, dt.$$

The mean quality-adjusted lifetime is then the sum of the utility of being in a transient state, u_p, times the expected time in the given state, Y_i. An example and further estimates can be found in Ref. 23.

8. Discussion

We have presented a flexible approach to the problem of censored data regression. The approach requires an (almost) unbiased estimator of the quantity to be modeled without covariates. The approach allows for censored data and truncated data, all that is needed is the unbiased estimator. Approach uses pseudo-observations from a jackknife like version of the statistic in generalized linear model.

We have presented a number of applications of the pseudo-observation approach. We assume that other key parameters that can be expressed as a mean function could have regression models developed by this approach.

References

1. D. R. Cox, *Journal of the Royal Statistical Society, Ser. B* **34** (1972).
2. P. K. Andersen and N. Keiding, *Statistical Methods in Medical Research* **11**, 91 (2002).
3. S. C. Cheng, J. P. Fine and L. J. Wei, *Biometrics* **54**, 219 (1998).
4. O. O. Aalen, *Statistics in Medicine* **8**, 907 (1989).
5. D. Y. Lin and Z. Ying, *The Annals of Statistics* **23**, 1712 (1995).
6. T. H. Scheike and M.-J. Zhang, *Scandinavian Journal of Statistics* **29**, 75 (2002).
7. J. P. Klein and M. Zhang, *Survival Anaysis*, in *Handbook of Statistics – Epidemiology and Medical Statistics*, eds. M. Rao and Roa (Elsevier, 2008), pp. 281–320.
8. P. K. Andersen, J. P. Klein and S. Rosthøj, *Biometrika* **90**, 15 (2003).

9. R. G. Miller, *Biometrika* **61**, 1 (1974).
10. B. Efron and R. Tibshirani, *An introduction to the bootstrap* (Chapman & Hall Ltd, 1993).
11. K.-Y. Liang and S. L. Zeger, *Biometrika* **73**, 13 (1986).
12. J. P. Klein and P. K. Andersen, *Biometrics* **61**, 223 (2005).
13. J. P. Klein, P. K. A. Andersen, B. L. Logan and G. Harhoff, *Statistics in Medicine* **26**, 4505 (2007).
14. B. Logan, J. Klein and M.-J. Zhang, *Biometrics* (in press).
15. J. P. Klein, M. Gerster, P. K. A. Andersen, S. Tarima and M. P. Perme, *Comput. Methods Programs Biomed.* **89**, 289 (2008).
16. E. J. Freireich, E. Gehan, E. Frei, L. R. Schroeder, I. J. Wolman, R. Anbari, E. O. Burgert, S. D. Mills, D. Pinkel, O. S. Selawry, J. H. Moon, B. R. Gendel, C. L. Spurr, R. Storrs, F. Haurani, B. Hoogstraten and S. Lee, *Blood* **21**, 699 (1963).
17. P. K. Andersen, M. G. Hansen and J. P. Klein, *Lifetime Data Analysis* **10**, 335 (2004).
18. J. P. Klein, *Statistics in Medicine* **25**, 1015 (2006).
19. P. K. Andersen and J. P. Klein, *Scandinavian Journal of Statistics* **34**, 3 (2007).
20. P. K. Andersen, S. Z. Abildstrom and S. Rosthøj, *Statistical Methods in Medical Research* **11**, 203 (2002).
21. P. K. Andersen, O. Borgan, R. D. Gill and N. Keiding, *Statistical Models Based on Counting Processes* (Springer-Verlag, 1993).
22. M. Pepe, *Journal of the American Statistical Association* **86**, 770 (1991).
23. G. Tunes-da Silva and J. P. Klein (2008), Submitted.

ON SOME NON-LINEAR RECURRENCES THAT ARISE IN COMPUTER SCIENCE

CHARLES KNESSL* and WOJCIECH SZPANKOWSKI†

Department of Mathematics, Statistics and Computer Science
University of Illinois at Chicago
Chicago, IL 60607-7045, USA
** E-mail: knessl@uic.edu*

Department of Computer Science, Purdue University
West Lafayette, IN 47907, USA
† E-mail: spa@cs.purdue.edu

We survey recent work on non-linear recurrence equations that arise in computer science or combinatorics. We consider the examples of the height of digital trees, the QUICKSORT algorithm, and enumeration problems concerning random binary trees characterized by nodes and path lengths. In each case a singular perturbation analysis of the recurrence yields insights into the asymptotic behavior, such as limiting distributions and tail probabilities.

1. Introduction

Many problems that arise in theoretical computer science lead naturally to solving non-linear recurrence or difference equations. These may sometimes be solved exactly but more frequently this is not the case, and an approximate analysis is necessary. Often the problems have a natural large parameter suggesting that an asymptotic analysis is appropriate. This large parameter could be the number of strings that are to be stored in a digital tree, or the number of items that are to be sorted by some algorithm. Some of the problems that arise are basic combinatorial enumeration problems.

Recent books that describe and analyze these types of problems are by Flajolet and Sedgewick [1] and Szpankowski [2].

In recent years we have analyzed some basic problems in computer science by using asymptotic methods of applied mathematics. These include methods for asymptotically evaluating sums and integrals, as well as perturbation methods such as matched asymptotics and WKB-type expansions. The latter are especially useful for difficult non-linear problems that cannot be solved explicitly.

In this note we survey some of the problems and the asymptotic solutions that we obtained. We consider digital trees, sorting algorithms, and enumeration problems that arise in studying numbers of nodes and paths in random binary trees.

2. Digital trees

Suppose that we have a set $\mathcal{S}$ of n strings, say $\mathcal{S} = \{s_1, s_2, \dots, s_n\}$ with each s_j being a finite or infinite sequence of 0's and 1's. Thus, e.g., $s_1 = 1\,0\,0\,1\,0\,1\,1\dots$. We assume a probabilistic model, namely that in a given position a 0 or 1 occurs with equal probability $= 1/2$, and that the positions are independent of one another. This is called a "symmetric Bernoulli model". We store the n strings in a tree. Different trees are obtained by using different rules for this storage. In a "trie" we store the string to the left or right of the root if $|\mathcal{S}| = 1$, according to whether the first symbol is a 1 or a 0. If $|\mathcal{S}| > 1$ we split the set of strings into two subsets according to whether the first symbol is a 1 or a 0. The trie is then built recursively and in Fig. 1 we illustrate this for an example with four strings. A second type of digital tree is a "PATRICIA trie", which can be obtained from a trie by eliminating nodes that have only one branch (see Fig. 1). A "digital search tree (DST)" is a further refinement that stores the strings in the internal nodes of the tree. To measure how efficiently a digital tree will store a large number n of strings we consider the "height" of the tree, which is defined as the largest path in the tree, and is related to the maximum search time. In the examples in Fig. 1 the heights are 3, 2, 2 for the respective cases of the trie, PATRICIA and DST.

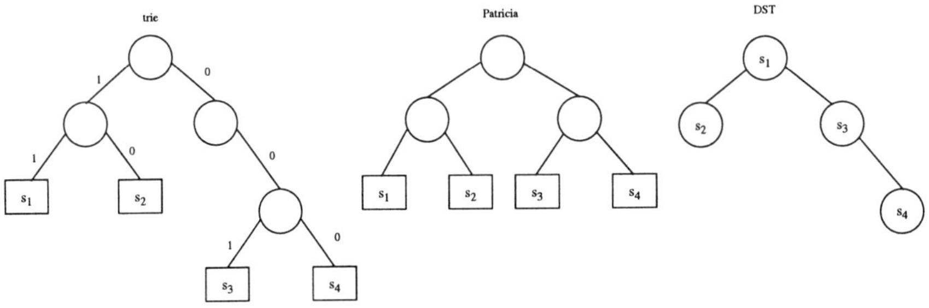

Fig. 1. A sketch of the three types of digital trees, as they each store the four strings $s_1 = 1\,1\,1\,0\,0\dots$, $s_2 = 1\,0\,1\,1\,1\dots$, $s_2 = 0\,0\,1\,1\,0\dots$ and $s_4 = 0\,0\,0\,0\,1\,1\dots$.

The height is a random variable whose probability distribution we denote by

$$h^{\mathrm{T}}(k,n) = \operatorname{Prob}\left[\mathcal{H}_n^{\mathrm{T}} \leqslant k\right] \tag{1}$$

where T is used to denote trie. Similarly we define the heights of PATRICIA and DST by $\mathcal{H}_n^{\mathrm{PAT}}$ and $\mathcal{H}_n^{\mathrm{DST}}$. The distribution functions satisfy the respective recurrences

$$h^{\mathrm{T}}(k+1,n) = 2^{-n}\sum_{i=0}^{n}\binom{n}{i} h^{\mathrm{T}}(k,i)h^{\mathrm{T}}(k,n-i),\ k \geqslant 0; \tag{2}$$
$$h^{\mathrm{T}}(0,0) = h^{\mathrm{T}}(0,1) = 1;\ h^{\mathrm{T}}(0,n) = 0,\ n \geqslant 2;$$

$$\begin{aligned} h^{\mathrm{PAT}}(k+1,n) &= 2^{1-n}h^{\mathrm{PAT}}(k+1,n) \\ &\quad + 2^{-n}\sum_{i=0}^{n-1}\binom{n}{i} h^{\mathrm{PAT}}(k,i)h^{\mathrm{PAT}}(k,n-i),\ n \geqslant 2,\ k \geqslant 0; \end{aligned} \tag{3}$$

$$h^{\mathrm{PAT}}(0,0) = h^{\mathrm{PAT}}(0,1) = 1;\ h^{\mathrm{PAT}}(0,n) = 0, n \geqslant 2;$$

$$h^{\mathrm{DST}}(k+1,n+1) = 2^{-n}\sum_{i=0}^{n} h^{\mathrm{DST}}(k,i)h^{\mathrm{DST}}(k,n-i),\ k \geqslant 0; \tag{4}$$
$$h^{\mathrm{DST}}(0,0) = h^{\mathrm{DST}}(0,1) = 1;\ h^{\mathrm{DST}}(0,n) = 0,\ n \geqslant 2.$$

These three recurrences look very similar but their solutions turn out to be much different. The case of tries can be explicitly solved and the mean height satisfies

$$E\left[\mathcal{H}_n^{\mathrm{T}}\right] \sim 2\log_2(n),\ n \to \infty. \tag{5}$$

The distribution follows a double exponential or extreme value law as $n \to \infty$ [3,4].

For the case of PATRICIA we analyzed Eq. (3) asymptotically as $n \to \infty$ using perturbation methods and found that $h^{\mathrm{PAT}}(k,n)$ behaves differently for the three scales (i) $k = n - O(1)$, (ii) $k = \log_2 n + O(1)$ and (iii) $2^k - n = O(1)$. Below we give some of our asymptotic formulas:

(i) $k = n - j,\ j = O(1),\ n \to \infty$

$$1 - h^{\mathrm{PAT}}(k,n) = \operatorname{Prob}\left[\mathcal{H}_n^{\mathrm{PAT}} > k\right] \sim 2^{-n^2/2}2^{(j-3/2)n} n!\rho_0 \mathrm{K}_j,$$

$$\mathrm{K}_j = 2^{-j^{-2}/2}2^{3j/2}\frac{1}{4\pi i}\oint z^{1-j}e^z \prod_{m=0}^{\infty}\left[\frac{1-\exp\left(-z2^{-m-1}\right)}{z2^{-m-1}}\right] dz$$

$$\rho_0 = \prod_{\ell=2}^{\infty}(1-2^{-\ell})^{-1} = 1.731\ldots.$$

(ii) $k, n \to \infty$; $\xi = n2^{-k}(1), 0 < \xi < 1$

$$h^{\mathrm{PAT}}(k,n) \sim \sqrt{1+2\xi\Phi'(\xi)+\xi^2\Phi''(\xi)}e^{-n\Phi(\xi)}$$

where $\Phi(\xi) > 0$ was determined numerically. Also, as $\xi \to 0^+$

$$\Phi(\xi) \sim \frac{1}{2}\rho_0 e^{\varphi(\log_2 \xi)}\xi^{3/2}\exp\left(-\frac{\log^2 \xi}{2\log 2}\right),$$

$$\varphi(x) = \frac{\log 2}{2}x(x+1) + \sum_{\ell=0}^{\infty}\log\left[\frac{1-\exp\left(-2^{x-\ell}\right)}{2^{x-\ell}}\right] + \sum_{\ell=1}^{\infty}\log\left[1-\exp\left(-2^{x+\ell}\right)\right],$$

and as $\xi \to 1^-$

$$\Phi(\xi) \sim D_1 + (1-\xi)\left[\log(1-\xi) - 1 - \log D_2\right],$$

$$D_1 = 1 + \log\left(\mathrm{K}_0^*\right), \quad D_2 = \mathrm{K}_1^*\mathrm{K}_0^*/e$$
$$\mathrm{K}_0^* = .6832\ldots, \quad \mathrm{K}_1^* = 1.259\ldots.$$

(iii) $k, n \to \infty$; $2^k - n = M = O(1)$

$$h_n^{\mathrm{PAT}}(k,n) \sim \frac{\sqrt{2\pi}}{M\,!}D_2^M n^{M+1/2}e^{-D_1 n}.$$

We can view (i) as the right tail and (iii) as the left tail of the distribution. Most of the probability mass is concentrated in that range of k where h^{PAT} changes from being ≈ 0 to being ≈ 1. We can show that this occurs in the asymptotic matching region between cases (i) and (ii). By using the behavior of $\Phi(\xi)$ as $\xi \to 0^+$ (which is in the matching region) we conclude that most probability mass occurs at the single point

$$k_1 = k_1(n) = 1 + \left\lfloor \log_2 n + \sqrt{2\log_2 n} - 3/2 \right\rfloor.$$

Thus $h^{\mathrm{PAT}}(k_1(n)-1,n) \approx 0$ while $h_n^{\mathrm{PAT}}(k_1(n),n) \approx 1$. This is true outside of very special subsequences of n, which lead to mass at exactly two points and which we precisely characterized in [5]. The mean thus satisfies

$$\mathrm{E}\left[\mathcal{H}_n^{\mathrm{PAT}}\right] = \log_2 n + \sqrt{2\log_2 n} + O(1), \tag{6}$$

which is smaller than the mean for tries (cf. Eq. (5)) by a factor of 1/2. Results similar to Eq. (6) were also obtained in [6,7] by probabilistic methods.

An analogous analysis of Eq. (4) for the DST model showed [8] that there are now four ranges of (k,n) that must be analyzed, which includes the three cases for PATRICIA and also the new scale where $k,n \to \infty$ with $k/n \in (0,1)$. We give below only our results for the mean

$$\mathrm{E}\left[\mathcal{H}_n^{\mathrm{DST}}\right] = \log_2 n + \sqrt{2\log_2 n} - \log_2\left(\sqrt{2\log_2 n}\right) + O(1), \qquad (7)$$

which shows that the DST height is typically smaller than the PATRICIA height by the $O(\log\log n)$ term. The first two terms in Eq. (7) were previously obtained in [9] by probability arguments, but this is not enough to distinguish DST from PATRICIA.

3. QUICKSORT algorithm

A popular sorting algorithm, that is taught in even elementary computer science classes, is QUICKSORT. This gives an efficient method of sorting n items. We assume that all possible $n!$ orderings of the items are equally likely and let $\mathcal{L}_n$ be the number of comparisons needed to sort completely the list.

In the algorithm we choose randomly one of the n items and then compare its rank to the remaining $(n-1)$ items. This divides the remaining items into two sublists, and these sublists are then sorted recursively by this method. It is well known [10] that as $n \to \infty$ the mean $\mathrm{E}[\mathcal{L}_n] \sim 2n\log n = O(n\log n)$, while the best case performance is $\sim n\log_2 n$, and the worst case is $\sim n^2/2$. Thus the mean is of the same order as the best case, with a larger constant since $2 > 1/\log 2$. Higher order moments are also readily computable but the full distribution of $\mathcal{L}_n$, $\Pr[\mathcal{L}_n = k]$, seems much harder. Its generating function

$$\mathrm{L}_n(u) = \sum_{k=0}^{\infty} \Pr\left[\mathcal{L}_n = k\right] u^k \qquad (8)$$

satisfies the non-linear recurrence

$$\mathrm{L}_{n+1}(u) = \frac{u^n}{n+1}\sum_{i=0}^{n} \mathrm{L}_i(u)\mathrm{L}_{n-i}(u), \; \mathrm{L}_0(u) = 1. \qquad (9)$$

We wish to study this equation for $n \to \infty$. We found that several different ranges of u lead to different expansions, but focus here only on one specific range of u, which will correspond to that range of k where most of the probability mass accumulates.

If we set $u = 1 + w/n$ (thus $u - 1 = O(n^{-1})$) with

$$\mathrm{L}_n(u) = e^{\mathrm{A}_n(u-1)}\mathrm{G}(n(u-1); n) = e^{\mathrm{A}_n w/n}\mathrm{G}(w; n), \tag{10}$$

$$\mathrm{A}_n = \mathrm{E}[\mathcal{L}_n] = 2(n+1)\left[\sum_{i=1}^{n} \frac{1}{i}\right] - 2n$$

then $\mathrm{G}(w; n) \to \mathrm{G}_0(w)$ as $w \to \infty$ where G_0 satisfies the non-linear integral equation

$$e^{-w}\mathrm{G}_0(x) = \int_0^1 e^{2\varphi(x)w}\mathrm{G}_0(wx)\mathrm{G}_0(w - wx)\, dx, \tag{11}$$

$$\varphi(x) = x\log x + (1-x)\log(1-x),\ \mathrm{G}_0(0) = 1,\ \mathrm{G}_0'(0) = 0.$$

Furthermore, $\Pr[\mathcal{L}_n - \mathrm{E}[\mathcal{L}_n] = ny] \sim n^{-1}P(y)$ where $P(y)$ satisfies the double integral equation

$$P(y+1) = \int_0^1 \int_{-\infty}^{\infty} P\left(xt + \frac{y - 2\varphi(x)}{2(1-x)}\right) \times P\left(-(1-x)t + \frac{y - 2\varphi(x)}{2(1-x)}\right) dt\, dx, \tag{12}$$

$$\int_{-\infty}^{\infty} P(y)\, dy = 1,\ \int_{-\infty}^{\infty} yP(y)\, dy = 0.$$

The functions G_0 and P are closely related, in fact $\mathrm{G}_0(w) = \int_{-\infty}^{\infty} e^{wy}P(y)\, dy$ is just the moment generating function of the continuous probability density $P(y)$.

An asymptotic analysis of Eqs. (11) and (12) yielded [11] the following results for $\mathrm{G}(w)$ as $w \to \pm\infty$:

$$\mathrm{G}_0(w) \sim \frac{2\sqrt{2}}{\sqrt{\pi \log 2}}\sqrt{-w}\exp\left[\left(\frac{1}{\log 2} - 2\right) w\log(-w) + \beta_0 w\right], \quad w \to -\infty \tag{13}$$

$$\mathrm{G}_0(w) \sim \frac{C_*}{w}e^{-w^2}e^{(1-2\gamma-2\log 2)w}\exp\left[\int_1^w \frac{2e^u}{u}\, du\right],\ w \to +\infty. \tag{14}$$

Here γ is the Euler constant, and β_0 and C_* are constants that must be evaluated numerically. The corresponding results for $P(y)$ as $y \to \pm\infty$ (the tails of the limiting QUICKSORT density) are

$$P(y) \sim \frac{1}{\pi e}\sqrt{\frac{2}{a}}\exp\left[\frac{\beta_0 - y}{a} - \frac{a}{e}\exp\left(\frac{\beta_0 - y}{a}\right)\right], \; y \to -\infty, \tag{15}$$
$$a = 2 - \frac{1}{\log 2},$$

$$P(y) \sim \frac{C_*}{\sqrt{8\pi}}\sqrt{\frac{y}{1 - 1/w_*}}e^{-w_*^2}e^{-(2\gamma + 2\log 2)w_*} \times \exp\left[-yw_* + \int_1^{w_*}\frac{2e^u}{u}\,du\right], \; y \to +\infty \tag{16}$$

where $w_* = w_*(y)$ is defined implicitly from

$$y = \frac{2}{w_*}e^{w_*};\; w_* \sim \log\left(\frac{y}{2}\right), y \to \infty. \tag{17}$$

It follows that the left tail of $P(y)$ is very thin (decaying as a double exponential) while the right tail is slightly thinner than an exponential (with $\log[P(y)] \sim -y\log y$ as $y \to +\infty$). Our results are much sharper than previous estimates in [12,13,14].

4. Binary trees

In Fig. 2 we sketch a typical binary tree with $n = 5$ nodes. Each node has an associated left and right path length. In going from the root of the tree to a given node we take a number of steps to the left and a number of steps to the right. The sum of these over all nodes is the right (resp. left) path length $\mathcal{R}$ (resp. $\mathcal{L}$) for the tree. The (total) path length is $\mathcal{P} = \mathcal{R} + \mathcal{L}$, which is the sum of the depths of the nodes in the tree. We then let $\mathcal{J} = \mathcal{R} - \mathcal{L}$ be the difference between the right and left path lengths.

We let $b(n,p)$ be the number of binary trees with n nodes and total path length p. The generating function $B_n(w) = \sum_{p=0}^{\infty} b(n,p)w^p$ satisfies

$$B_{n+1}(w) = w^n \sum_{k=0}^{n} B_k(w)B_{n-k}(w), \; n \geqslant 0 \tag{18}$$

with $B_0(w) = 1$. It was previously established that $B_n(1)$, the total number of trees with n nodes, is the Catalan number $\mathcal{C}_n = \frac{1}{n+1}\binom{2n}{n}$. Also, the fraction of trees with n nodes that have path length p, $b(n,p)/\sum_{p=0}^{\infty} b(n,p)$,

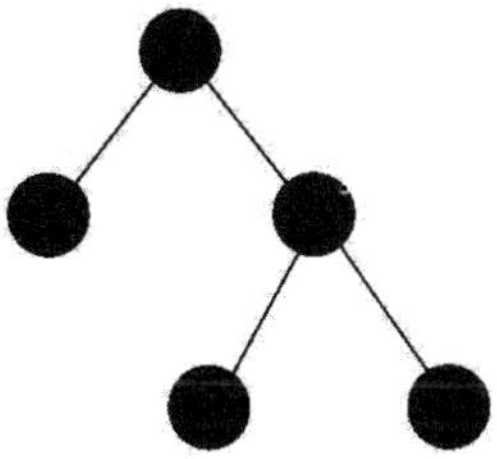

Fig. 2. A sketch of a binary tree with 5 nodes, total path length $\mathcal{P} = 6$, right path length $\mathcal{R} = 4$, and left path length $\mathcal{L} = 2$ (thus $\mathcal{J} = 2$).

follows an Airy distribution [15] with the scaling $p = O(n^{3/2})$. A more difficult problem is to fix the path length p and then study the distribution of the number of nodes, i.e., $b(n,p)/\sum_{n=0}^{\infty} b(n,p)$, or to study the double sequence $b(n,p)$. In [16] we analyzed $b(n,p)$ for the following scales: (i) $p, n \to \infty$ with $p = \binom{n}{2} - O(1)$, (ii) $p = O(n^2)$, (iii) $p = O(n^{3/2})$ (leading to leading order to the Airy distribution), (iv) $p = O(n^{4/3})$, and (v) $p = n \log_2 n + O(n)$. Having a thorough understanding of $b(n,p)$ in all of the different ranges allowed us to first estimate the number of trees (regardless of the number of nodes) that have path length p. Its exponential growth rate takes the form

$$\log\left[\sum_{n=0}^{\infty} b(n,p)\right] = \frac{2p\log^2 2}{\log p}\left[1 - C_0(\log p)^{-2/3} + O\left((\log p)^{-1}\right)\right] \tag{19}$$

where $p \to \infty$ and $C_0 = (2\log 2)^{1/3}|r_0|$, where $r_0 = \max\{z\colon Ai(z) = 0\} = -2.3381\ldots$ is the maximal root of the Airy function. The leading term in (19) was also obtained using combinatorial arguments by Seroussi [17]. To understand the fraction of trees that have n nodes for a fixed path length p we must analyze the asymptotic matching region between cases (iv) and (v) above. Examining carefully the scale $p = n\log_2 n + O[n(\log n)^{1/3}]$ we obtained the Gaussian limit law

$$\frac{b(n,p)}{\sum_{n=0}^{\infty} b(n,p)} \approx \frac{1}{\sqrt{2\pi\mathcal{V}(p)}} \exp\left[-\frac{(n-\mathcal{N}(p))^2}{2\mathcal{V}(p)}\right] \tag{20}$$

where the mean and variance are

$$\mathcal{N}(p) = \frac{p\log 2}{\log p}\left[1 - \frac{2^{4/3}}{3}(\log 2)^{1/3}\frac{|r_0|}{(\log p)^{2/3}} + O\left((\log p)^{-1}\right)\right] \tag{21}$$

$$\mathcal{V}(p) \sim \frac{p}{(\log p)^{5/3}}\frac{2^{1/3}}{9}(\log 2)^{1/3}|r_0|.$$

We next examine binary trees, but now distinguish between the left and right paths. We let $b(n,r,\ell)$ be the number of such trees with right (resp., left) path $= r$ (resp., ℓ). Its double generating function $G_n(w,v) = \sum_{r=0}^{\infty}\sum_{\ell=0}^{\infty} b(n,r,\ell)w^r v^\ell$ satisfies

$$G_{n+1}(w,v) = \sum_{k=0}^{\infty} w^k v^{n-k} G_k(w,v) G_{n-k}(w,v) \tag{22}$$

with $G_0(w,v) = 1$. In [18] we analyzed the joint left–right path distribution for $n \to \infty$ with the scaling $\mathcal{P}(n^{3/2})$ and $\mathcal{J}(n^{5/4})$. Here we only discuss the path length difference, whose generating function is $G_n(w,w^{-1})$. The fraction of trees whose path length difference is $\mathcal{J} = n^{5/4}\beta = O(n^{5/4})$ satisfies the limit law

$$\frac{1}{\mathcal{C}_n}\sum_{\ell} b(n,\ell+\mathcal{J},\ell) \sim n^{-5/4} p_-(\beta) \tag{23}$$

as $n \to \infty$, where $p_-(\beta)$ is a continuous density that satisfies $p_-(\beta) = p_-(-\beta)$, $\int_{-\infty}^{\infty} p_-(\beta)\,d\beta = 1$ and has the properties

$$p_-(\beta) \sim \sqrt{\frac{5}{6}}(5\beta)^{1/3}\widetilde{C}\exp\left[-\frac{3}{4}5^{1/3}\beta^{4/3}\right],\ \beta \to \infty,$$
$$\widetilde{C} = .5513\ldots,$$
$$p_-(0) = .4572\ldots,\ p_-''(0) = -.7146\ldots.$$

Also, $p_-(\beta)$ has a unique inflection point for $\beta > 0$, at $\beta = .7589\ldots$. Other recent investigations into the path length difference and its asymptotic properties appear in [19,20]. We also show in [18] that the moment generations function of p_- is

$$\int_{-\infty}^{\infty} e^{\beta\theta} p_-(\beta)\,d\beta = 1 + \sqrt{\pi}\,\overline{\mathrm{H}}(\theta)$$

where $\overline{\mathrm{H}}(\theta) = \theta^{6/5}\Delta(\theta^{4/5}) = B^{3/2}\Delta(B)$ $(B = \theta^{4/5})$ and $\Delta(B)$ satisfies the non-linear integral equation

$$\int_0^B \Delta(\xi)\Delta(B-\xi)\,d\xi + 2B^2\Delta(B) + 2\sqrt{\frac{B}{\pi}} = \frac{4}{\sqrt{\pi}}\int_0^B \frac{\Delta'(\xi)}{\sqrt{B-\xi}}\,d\xi. \tag{24}$$

This characterizes the moment generating function for θ real and positive. For θ purely imaginary we let $\theta = ix$, $y = x^{4/5}$, $\overline{\mathrm{H}}(ix) = -y^{3/2}\Lambda(y)$ and $\mathrm{U}(\varphi) = \int_0^\infty e^{-y\theta}\Lambda(y)\,dy$. Then $\mathrm{U}(\varphi)$ satisfies the non-linear ODE

$$2\mathrm{U}''(\varphi) + \mathrm{U}^2(\varphi) + 4\sqrt{\varphi}\,\mathrm{U}(\varphi) = \varphi^{-3/2}, \tag{25}$$

which is closely related to the first Painlevè transcendent.

Acknowledgments

The work of C.K. was partially supported by NSF grant DMS-0503745; the work of W.S. by NSF grants DMS-0503742, DMS-0800568, and CCF-0513636, NSA grant H 98230-08-1-0092, NIH grant R01 GM068959-01, and AFOSR grant FA8655-07-1-3071.

References

1. P. Flajolet and R. Sedgewick, *Analytic Combinatorics* (Cambridge University Press, 2008).
2. W. Szpankowski, *Average Case Analysis of Algorithms and Sequences* (Wiley–Interscience, New York, 2001).
3. P. Flajolet, *Acta Inform.* **20**, 345 (1983).
4. P. Jacquet and M. Règnier, *Trie partitioning process: Limiting distributions*, Lecture Notes in Comp. Sci., Vol. 214 (Springer, New York, 1986).
5. C. Knessl and W. Szpankowski, *J. Algorithms* **44**, 63 (2002).
6. L. Devroye, *Random Structures Algorithms* **3**, 203 (1992).
7. B. Pittel and H. Rubin, *J. Combin. Theory Ser. A* **55**, 292 (1990).
8. C. Knessl and W. Szpankowski, *SIAM J. Comput.* **30**, 923 (2000).
9. D. Aldous and P. Shields, *Probab. Theory Related Fields* **79**, 509 (1988).
10. D. Knuth, *The Art of Computer Programming. Sorting and Searching*, 2nd edn. (Addison–Wesley, Reading, MA, 1998).
11. C. Knessl and W. Szpankowski, *Discrete Math. Theor. Comp. Sci.* **3**, 43 (1999).
12. P. Hennequin, *Theor. Inform. Appl.* **23**, 317 (1989).
13. U. Rösler, *Theor. Inform. Appl.* **25**, 85 (1991).
14. C. J. McDiarmid and R. Hayward, *J. Algorithms* **21**, 476 (1996).
15. P. Flajolet and B. Louchard, *Algorithmica* **31**, 361 (2001).
16. C. Knessl and W. Szpankowski, *Discrete Math. Theor. Comp. Sci.* **7**, 313 (2005).
17. G. Seroussi, *Algorithmica* **46**, 557 (2006).
18. C. Knessl and W. Szpankowski, *Studies Appl. Math.* **117**, 109 (2006).
19. J.-F. Marckert, *Random Structure Algorithms* **24**, 118 (2004).
20. S. Janson, *Algorithmica* **46**, 419 (2006).

SMALL-SAMPLE INFERENCE FOR NON-INFERIORITY IN BINOMIAL EXPERIMENTS

JUDY DAVIDSON and JOHN KOLASSA

Department of Statistics and Biostatistics Rutgers University,
Piscataway, NJ 08855, USA
**E-mail: kolassa@stat.rutgers.edu*

We compare the exact sizes and powers of various procedures for testing hypotheses concerning differences of sample proportions. We eliminate the effect of the remaining parameter following the method of Berger and Boos,[1] who suggest constructing a confidence interval for the nuisance parameter, calculating the supremum of the p-value as the nuisance parameter varies over this interval, and reporting as the p-value for the test this supremum, plus the complement of the coverage probability for the interval. Our method forces the size, i.e. the power at the null hypothesis, to be strictly controlled, compared to the standard z-test, which is anti-conservative. We also found that we can make modest improvements in power by optimizing over the coverage probability of the nuisance parameter confidence interval.

Keywords: exact inference

1. Introduction

A commonly–used hypothesis test involves taking the difference in proportions for two independent samples. The proportions are binomial proportions that represent successes in different trials. A practical reason for doing this is to see if a new medication is just as good or better as one that is already used. This is called non-inferiority or equivalence testing. But there is a problem with this type of hypothesis testing. There is no test statistic whose distribution depends only on the difference of these two proportions, so we have to introduce a nuisance parameter. Practitioners typically address the dependence of the test statistic distribution on this nuisance parameter by estimating it and treating the estimated value as the true value.

One might measure proportion differences using the difference, relative risk, and odds ratio. When the null hypothesis is expressed in terms of the odds ratio, one might condition on the overall number of successes to remove the effect of the nuisance parameter. This conditioning often injects unwanted conservatism, and is unavailable when the null hypothesis is expressed in terms of proportion differences. Newcombe[2] surveys eleven methods for the complementary problem of confidence interval construction for proportion differences.

Berger and Boos[1] propose eliminating the effect of the nuisance parameter by calculating a confidence interval for this parameter with a high coverage probability, maximizing the resulting p-value over values of the nuisance parameter in the interval, and adding the complement of the coverage probability to the resulting maximum p-value. Berger[3] discusses the properties of this test further. Freidlin and Gastwirth[4] extend this technique to sets of pairs of binomial observations. Berger and Sidik[5] apply similar techniques to matched-pairs designs.

When sample sizes are large, one typically instead reports the p-value corresponding to an estimated value of the nuisance parameter. A problem occurs when the sample size is small, since in this case the modified test statistic may reject the null hypothesis more often than expected under the null hypothesis. This is a significant problem, since it is not always feasible to have a large sample. When testing a new treatment, there are not necessarily going to be plentiful participants. Large-scale samples also have much higher monetary costs and can be inefficient.

Using the maximum p-value is a problem because in some cases, for most values of the nuisance parameter, the p-value is significant, but for a small number, the p-value is insignificant. When we take the max, we get one that is insignificant, and therefore cannot reject the null hypothesis. One way of dealing with this is by taking the maximum p-value over a confidence interval for the nuisance parameter (Berger and Boos, 1994, p. 1013). Then we add the type I error to this p-value to account for the possible error introduced by taking the confidence interval.

The current methods do not account for hypothesis testing where we want to test the null hypothesis that the difference in the proportions is non-zero. This is not trivial because we may want to market a drug if it has success proportion within .1 (or some other fraction) of what is currently used since it might have other benefits such as being safer, easier, and cheaper (Chan 37). The power of the tests is another important factor that needs to be included when analyzing results of different methods.

2. Notation

Denote the first sample size by m, the first number of successes by X, and the success proportion by π. Denote the second sample size by n, the second number of successes by Y, and the success proportion by ρ. We are testing the null hypothesis $H_0 : \pi = \rho + \delta_0$ versus the alternative hypothesis $H_A : \pi > \rho + \delta_A$. Roehmeland Mansmann[6] consider more complicated non-equality null hypotheses. We use the test statistic

$$Z = (X/m - Y/n - \delta_0)/\sqrt{\hat{\pi}(1-\hat{\pi})/m + \hat{\rho} \times (1-\hat{\rho})/n},$$

for $\hat{\pi}$ and $\hat{\rho}$ the maximum likelihood estimator for π and ρ under the restriction $\pi - \rho = \delta_0$. Then $\hat{\pi}$ solves a cubic equation. When $\delta = 0$, then $\hat{\pi} = \hat{\rho} = (X+Y)/(n+m)$. Mehrotra, Chan, and Berger[7] note that this choice is preferable to $\hat{\pi} = X/m$ and $\hat{\rho} = Y/n$. Wallenstein[8] discusses other formulas for S for use in confidence intervals for δ.

Certain extreme cases must be accounted for. When $X = m$ and $Y = n$, or $X = 0$ and $Y = 0$, no information exists for rejecting the null hypothesis. The p-value is set to 1. When $X = m$ and $Y = 0$, this is good evidence to reject the null hypothesis. The standard error S is 0 for this Y and X, and set Z to a very large constant. When $X = 0$ and $Y = n$, we should not reject the null hypothesis. Again, Z is undefined, so we set it to be a very large negative constant.

The sampling distribution of Z depends on $\eta = (\pi + \rho)/2$. Following Berger and Boos,[1] we create a confidence interval $\mathcal{H}(\gamma)$ for η, with confidence level $1 - \gamma$, and present as the p-value $\varpi_\gamma(x, y) = \max_{\eta \in \mathcal{H}(\gamma)} \mathrm{P}\left[Z \geq z(x, y)\right]$.

When the equality in the definition of H_0 is replaced by an inequality, $\varpi_\gamma(x, y)$ remains a valid p-value for each γ. Roehmeland Mansmann[9] and Chan[10] discuss this issue. Roehmel[11] presents a proof of the validity of ϖ_γ as a p-value, but notes that it lacks a certain desirable monotonicity property; that is, one expects that $X = x, Y = y$ presents at least as much evidence against H_0 as $X = x - 1, Y = y$ and $X = x, Y = y + 1$, and so one would expect that $\varpi_\gamma(x, y) \leq \varpi_\gamma(x - 1, y)$ and $\varpi_\gamma(x, y) \leq \varpi_\gamma(x, y + 1)$ for all x and y, but Roehmel[11] shows that this need not hold.

3. Results

The critical region of the test of nominal level α associated with the p-value ϖ_γ is $\mathcal{C}(\gamma, \alpha) = \{(x, y) | x \in \{0, \ldots, m\}, y \in \{0, \ldots, n\}, \varpi_\gamma(x, y) < \alpha\}$, and the associated power is

$$\varrho(\gamma, \alpha, \delta, \eta) = \sum_{(x,y)\in\mathcal{C}(\gamma,\alpha)} \varpi$$

for

$$\varpi = \binom{m}{x}\binom{n}{y}(\eta + \delta/2)^x(1-\eta-\delta/2)^{m-x}(\eta-\delta/2)^y(1-\eta+\delta/2)^{n-y},$$

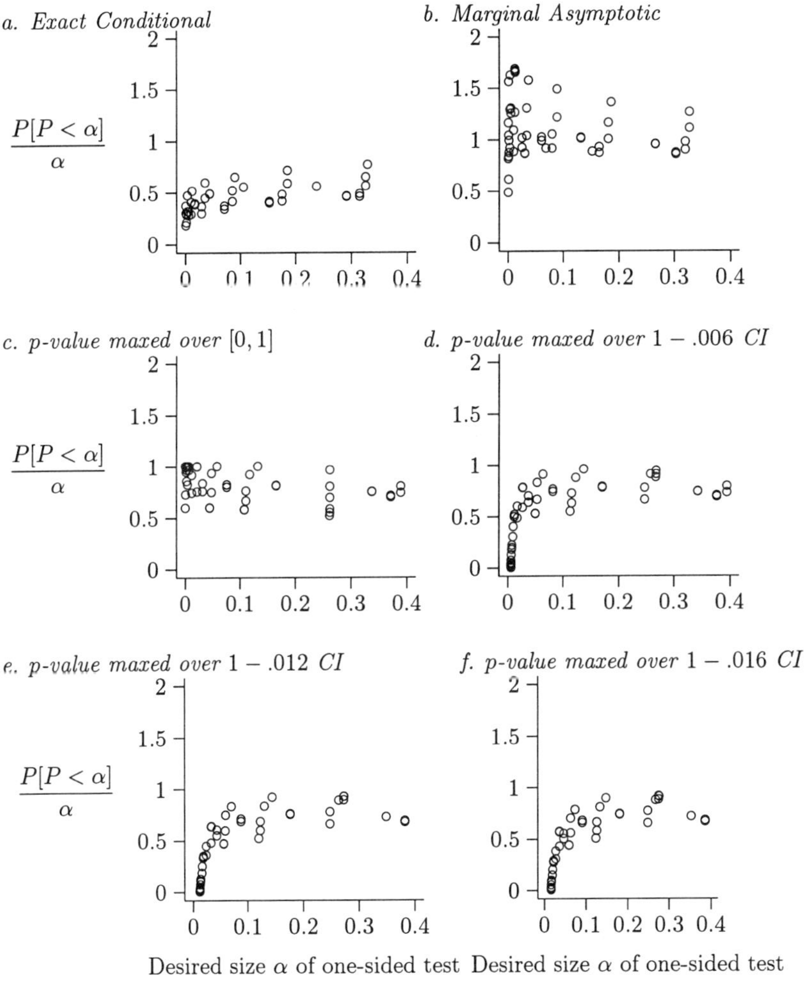

Fig. 1. Size of Tests of Equal Proportions, Samples of Size 10.

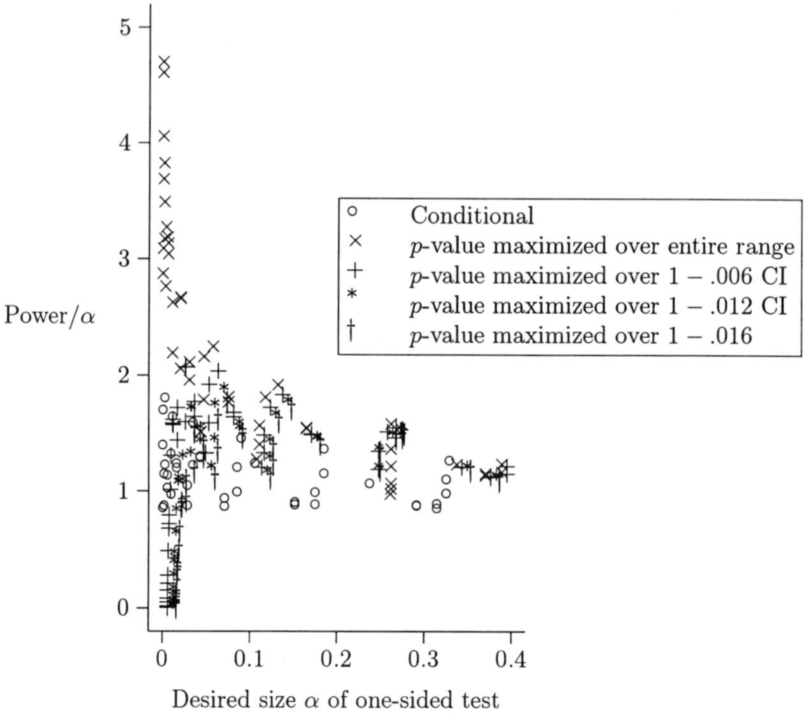

Fig. 2. Power of Tests of Equal Proportions, Samples of Size 10.

the p-values that fell into the critical region. We compared it with the standard z-test method, and also with the conditional method that uses the odds ratio and hypergeometric distribution.

Figure 1 shows actual size, with $\delta_0 = 0$, for the asymptotic test, the conditional test, the test that maximizes the p-value over all values of the nuisance parameter, and the test that maximizes the p-value over intervals of varying confidence for the nuisance parameter, and augmented by the complement of the confidence level, as proposed by Berger and Boos.[1] Note that the p-value maximized over all value of the nuisance parameter is a special case of the Berger and Boos[1] p-value, in which the interval has confidence 1. In this case, both binomial samples have size 10. A violation of type I error control is indicated in Fig. 1 by a value above 1. The asymptotic tests clearly do not control type I error. Conservativeness is indicated in Fig. 1 by values below 0. The conditional test is most conservative. Now consider tests maximized over values of the nuisance parameter. Tests of small size

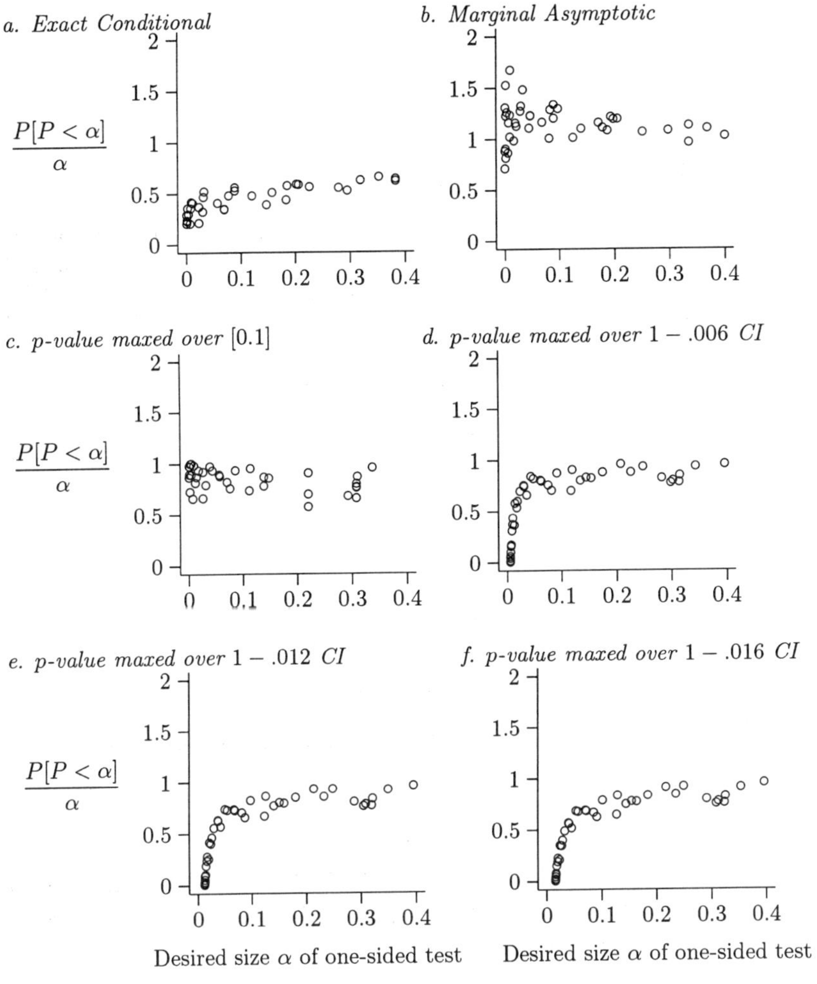

Fig. 3. Size of Tests of Equal Proportions, Samples of Size 10 and 8.

are least conservative when the confidence level is high, and tests of larger size are least conservative when the confidence level declines somewhat. In each case, $\pi = \rho = .5$.

Figure 2 shows the power for the various tests at $\delta_A = 0.1$. Here $(\pi + \rho)/2 = .5$. Figures 3 and 4 show the analogous pictures when the binomial sample sizes are 10 and 8.

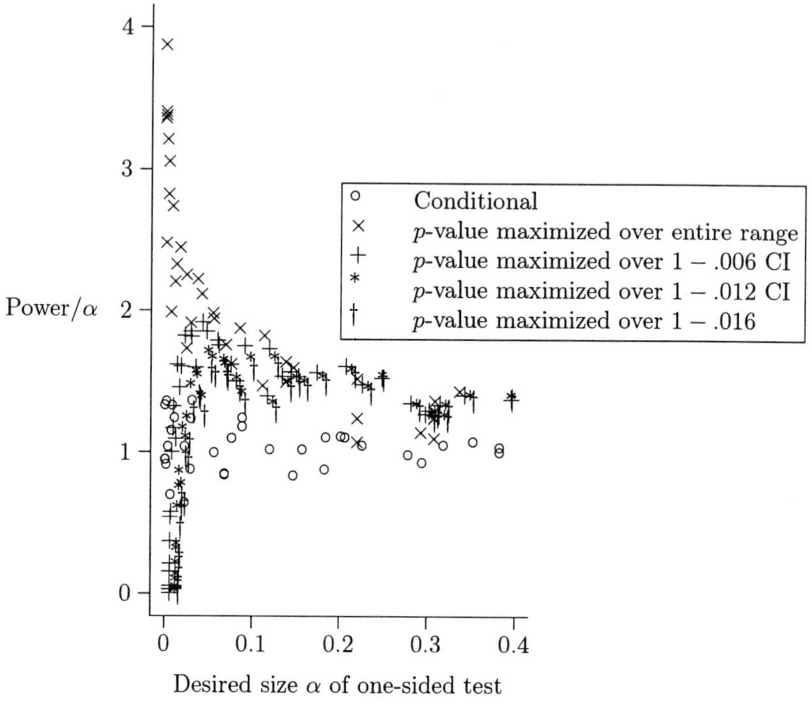

Fig. 4. Power of Tests of Equal Proportions, Samples of Size 10 and 8.

Acknowledgments

This research was funded by NSF grant DMS 0505499, and by REU award 0138973 from the NSF to Center for Discrete Mathematics and Theoretical Computer Science at Rutgers University.

References

1. R. Berger and D. Boos, *Journal of the American Statistical Association* **89**, 1012 (1994).
2. R. Newcombe, *Statistics in Medicine* **17**, 873 (1998).
3. R. Berger, *American Statistician* **50**, 314 (1996).
4. B. Freidlin and J. Gastwirth, *Biometrics* **55**, 264 (1999).
5. R. Berger and K. Sidik, *Statistial Methods in Medical Research* **12**, 91 (2003).
6. J. Röhmel and U. Mansmann, *Biometrical Journal* **41**, 149 (1999b).
7. C. I. Mehrotra, D.V. and R. Berger, *Biometrics* **59**, 441 (2003).
8. S. Wallenstein, *Statistics in Medicine* **18**, 1329 (1997).

9. J. Röhmel and U. Mansmann, *Statistis in Medicine* **18**, 1734 (1999a).
10. I. Chan, *Statistis in Medicine* **18**, 1735 (1999).
11. J. Röhmel, *Biometrical Journal* **47**, 37 (2005).

TERMINATION OF CARDIAC REENTRY

T. KROGH-MADSEN and D. J. CHRISTINI*

Department of Medicine, Division of Cardiology, Weill Cornell Medical College, New York, New York 10021, USA
** E-mail: dchristi@med.cornell.edu*
www.med.cornell.edu

Implantable cardioverter defibrillators (ICDs) have become standard therapy for many patients at risk for reentrant ventricular tachycardias. By applying one or more series of suprathreshold stimuli, the antitachycardia pacing modality of ICDs successfully terminates ventricular tachycardia in up to 90% of attempted trials. However, many aspects of why antitachycardia pacing is successful are unknown. We conducted numerical simulations to investigate the dynamics of reentry termination, and have in particular studied situations where (1) the stimulus site is located at some distance away from the reentrant loop, and (2) the reentrant circuit is sufficiently short that a period-doubling bifurcation to repolarization alternans has occurred. Our study shows that when applying a single stimulus, termination of reentry is possible only when the stimulus site is very close to the reentrant loop. However, during alternans, termination is much facilitated by the presence of spatial gradients in action potential duration and recovery time.

Keywords: Cardiac reentry; Termination; Alternans; Pacing.

1. Introduction

Cardiac reentry occurs when tissue is activated repeatedly by one or more waves that again and again reenters the same anatomical region. Such reentry can cause excessively rapid activation and contraction of the heart, potentially causing a fatal reduction in the efficiency with which blood is pumped. Because of this risk, termination of reentry is critically important.

Patients at risk for cardiac reentry may receive an implantable cardioverter defibrillator (ICD). In the case of a single reentrant wave occurring in the main pumping chambers of the heart (this is associated with the cardiac arrhythmia called ventricular tachycardia), the ICD attempts termination by delivering a series of suprathreshold electrical stimuli. This is called antitachycardia pacing.

Despite its high success rate (up to 90%), many aspects of how antitachycardia pacing works are not known. One part which is known, however, is how a single (very) well-timed stimulus may terminate a reentrant wave. This occurs through a mechanism called unidirectional block, in which the stimulus induces a wave that travels only in the direction retrograde (i.e., reverse) to the original reentrant wave because the tissue in the anterograde (i.e., forward) direction is still refractory. This retrograde wave and the original reentrant wave then collide and annihilate each other at the antipodal point on the ring.

If the stimulus is delivered too late, it causes resetting of the reentry rather than termination. Hence, the effect of a single stimulus may be described quantitatively by a phase resetting curve.[1] However, iterating the phase resetting curve does not predict the effects of multiple, rapidly induced stimuli (the ICD typically delivers 8 to 12 stimuli) very well because the state-point does not return to the limit cycle between the rapid stimulations.[2] Hence, there is no appropriate theoretical framework for understanding the effects of such stimulus series. On the other hand, the dynamics resulting on a short ring, where each virtual cell is stimulated repeatedly and rapidly by the reentrant wave, is well-known.[3] The extent to which the termination dynamics in this situation are comparable to those arising due to rapid point-stimulation on a longer ring is not known.

Another issue is that in the heart, the stimulating electrode is not necessarily situated within the reentrant loop. Hence, waves originating at the ICD electrode must travel to the reentrant loop before being able to perturb the rhythm. However, the reentrant wave and its refractory wake limit such propagation. It is thought that successive stimuli "peel back" refractoriness more and more, allowing the stimulus-induced waves to approach and eventually enter the reentrant circuit, however, there is no clear physical mechanism explaining this.

As a first step towards elucidating the mechanisms of antitachycardia pacing, we have investigated (1) the effects of the stimulus site being located at some distance away from the reentrant loop and (2) the dynamics of single-pulse termination on a short ring.

2. Termination of reentry in a loop-and-tail model

2.1. *Methods*

Simulations were carried out using the Aliev-Panfilov version of the FitzHugh-Nagumo model,[4] which is modified for modeling cardiac tissue.

The model equations are:

$$\frac{\partial u}{\partial t} = D\frac{\partial^2 u}{\partial x^2} - ku(u-a)(u-1) - uv + I, \quad \frac{\partial v}{\partial t} = \epsilon(u,v)(-v - ku(u-a-1)), \quad (1)$$

with t measured in time units (t.u.) and x in space units (s.u.), and where $\epsilon(u,v) = \epsilon_0 + \mu_1 v/(\mu_2 + u)$, $\epsilon_0 = 0.002$, $\mu_1 = 0.2$, $\mu_2 = 0.3$, $k = 8.0$, $a = 0.15$, $D = 1$ s.u.^{2}/t.u., and I is the injected stimulus current (amplitude 30, duration 0.08 t.u., at a distance x_s away from the ring, at time t_c following the previous action potential upstroke).

For numerical integration we used a finite-difference method with forward Euler. The sizes of the spatial and temporal steps were $dx = 0.5$ s.u. and $dt = 0.02$ t.u. The loop was 100 s.u. long, as was the tail. The stimulus site was varied systematically between different locations on the tail.

2.2. *Results*

First, we placed the stimulus site very close to the reentrant loop, since one would expect (by continuation) this case to be similar to the well-known case of no tail. Indeed, Fig. 1 shows this to be the case: if delivered early (t_c = 21.1 t.u.; panel A), the stimulus falls within the refractory period and has no effect on the dynamics. A stimulus delivered after the refractory period

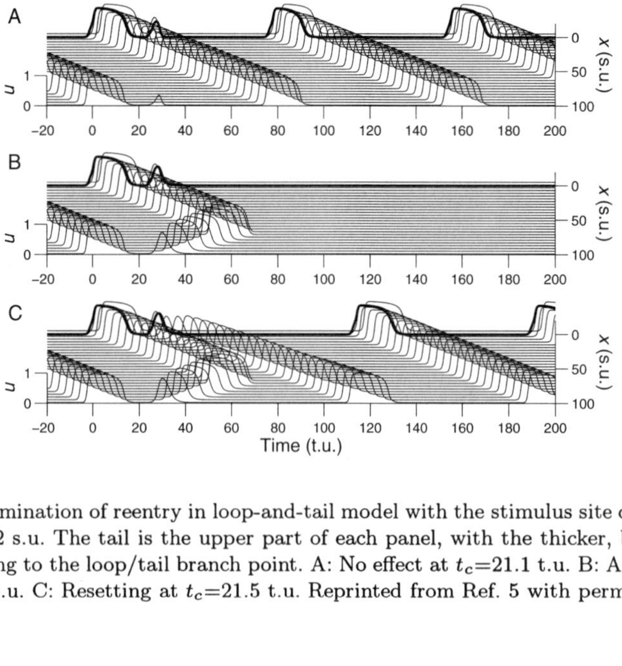

Fig. 1. Termination of reentry in loop-and-tail model with the stimulus site close to the loop, $x_s = 2$ s.u. The tail is the upper part of each panel, with the thicker, black trace corresponding to the loop/tail branch point. A: No effect at t_c=21.1 t.u. B: Annihilation at t_c=21.3 t.u. C: Resetting at t_c=21.5 t.u. Reprinted from Ref. 5 with permission.

(t_c=21.5 t.u.; panel C) induces a wave which travels in both directions on the ring, with the retrograde wave annihilating the original reentrant wave while the anterograde wave keeps circulating, thus maintaining the reentry albeit at a reset phase. However, a well-timed stimulus (t_c=21.3 t.u.; panel B) induces a wave which travels down to the ring, where it propagates in the retrograde direction only, since the tissue there has had more time to recover than the tissue in the anterograde direction.

The time interval in which a stimulus leads to reentry termination is termed the vulnerable window (VW). Introducing a tail changes VW quantitatively. While it is 2.9 t.u. in the case of $x_s = 0$ s.u. (i.e., when the stimulus site is on the ring), it is only 0.2 t.u. when $x_s = 2$ s.u. (Fig. 2A). Indeed, VW decreases linearly for small x_s and has dropped to within one integration time step (0.02 t.u.) at $x_s = 3.5$ s.u. The linear fall-off is consistent with the mechanism illustrated in Fig. 2B: the VW is decreased by the time it takes the reentrant wave to travel up the tail to the stimulus site plus the time it takes the stimulus-induced wave to propagate down to the ring. Hence, assuming constant conduction velocity (CV), the VW varies as

$$\mathrm{VW}(x_s) = \mathrm{VW}_0 - 2x_s/\mathrm{CV}, \tag{2}$$

where VW_0 is the VW at location $x_s = 0$. This result (shown as the dashed line in Fig. 2B) agrees very well with the simulation results.

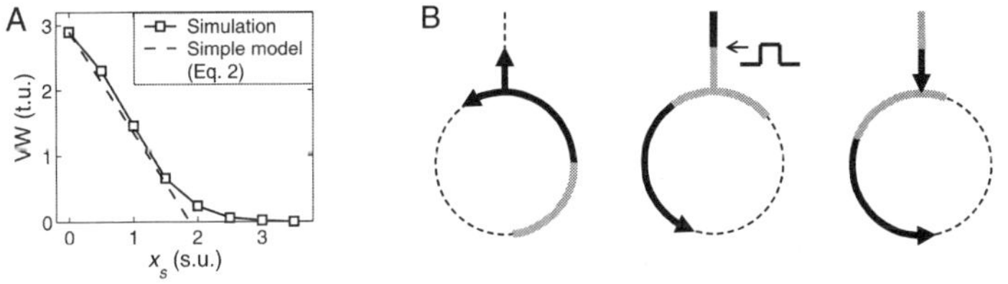

Fig. 2. A: Drop in VW for increasing distance, x_s, of the stimulus site from the loop. B: Reentrant wave (*thick, black curve*) rotating in loop-and-tail model, followed by VW (*gray*). The earliest time at which a stimulus delivered on the tail will be able to propagate down to the loop is at the start of the VW (middle panel). For this wave to terminate the reentry, it must block unidirectionally on the loop, i.e., it must reach the loop within the VW (right panel). Panel A is reprinted from Ref. 5 with permission.

2.3. *Conclusions*

Our main point for this part is that for a single stimulus in a loop-and-tail geometry, termination of reentry can occur only when the stimulus site is located closer to the loop than a critical distance, x_c, given from Eq. 2 by $x_c = \mathrm{VW}_0\mathrm{CV}/2$. Using physical values ($\mathrm{VW}_0 \approx 1.5$ ms, CV ≈ 50 cm/s), we estimate $x_c \approx 0.04$ cm. Hence, for termination to work with a realistic (i.e., cm scale) distance of the stimulus electrode from the loop, the VW must be increased. One way to do this may be by rapid pacing or rapid intrinsic activation, which is what we address in the next part.

3. Termination of reentry in a short loop model

3.1. *Methods*

Because we expect the quantitative features of reentry termination to depend on the repolarization dynamics, we opted for a more realistic ionic model for this part of the study. Hence, we used a recent Hodgkin–Huxley-type model of canine ventricular cells;[6] we do not print the equations here for brevity. The model was integrated numerically using a finite-difference method with forward Euler, where $dx = 0.015$ cm and $dt = 0.01$ ms.

3.2. *Results*

When paced at a rapid frequency, cardiac cells typically undergo a period-doubling bifurcation to alternans, where a long action potential is followed by a short one, and so forth.[7] Such alternans also occurs for reentry around a loop, when the loop is sufficiently short that each region is stimulated rapidly by the reentrant wave.[3,8]

An example of alternans in a 10 cm loop is shown in Fig. 3A. In addition to alternans in action potential duration (APD), the recovery interval (termed the diastolic interval; DI) also alternates. Due to dispersion, there is also spatial variation in both APD and DI, even though the model is homogenous.[9,10] Thus, at $x = 0$ cm, there is very small amplitude alternans (Fig. 3B), while at $x = 4.5$ cm the alternans is of large amplitude (Fig. 3C).

Because of the presence of these spatiotemporal gradients, the effects of a single stimulus depend on both its timing and location. In Tbl. 1, we summarize the results of applying a single stimulus at one of the two locations shown in Fig. 3, following either the long (L) or the short (S) action potential. In some of these cases the VW is much increased from its value of 1.6 ms (due to unidirectional block) in a longer ring (16 cm) without alternans.

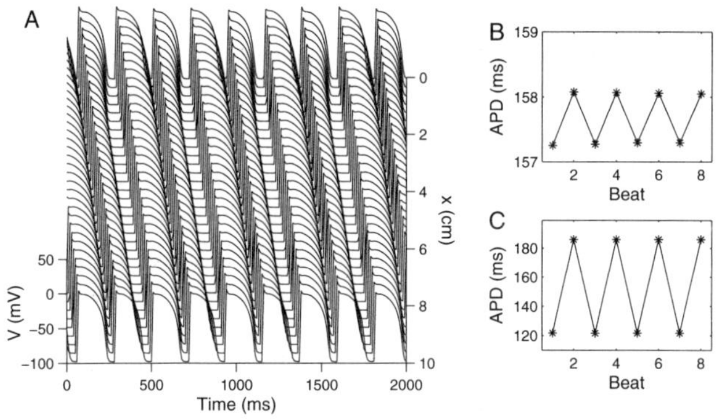

Fig. 3. A: Alternans and spatial heterogeneity on short ring. B: Small alternans amplitude at $x = 0$ cm. C: Large alternans amplitude at $x = 4.5$ cm.

Table 1. VW for termination of reentry on a short ring.

Stimulus site (cm)	Stimulus follows	VW unidir. block (ms)	VW $A \rightarrow F$ (ms)	VW $A \rightarrow A$ (ms)	VW total (ms)
0	S	< 1	< 1	< 1	< 1
0	L	**15**	< 1	**8**	**24**
4.5	L	1	< 1	< 1	< 1
4.5	S	2	**50**	**5**	**58**
Examples		Fig. 4A	Fig. 4C	Fig. 4B	

As indicated in the table, the mechanism of termination depends on stimulus timing and APD dynamics. For example, when $x = 0$ cm, and the stimulus follows the (slightly) longer action potential, there is a large VW (15 ms) for unidirectional block. This is due to the spatial gradient in DI as shown in Fig. 4A: the anterograde wave ("A") meets tissue which has decreasing DI and hence blocks easily, while the retrograde wave "R" meets tissue with increasing DI and hence propagates easily until it and the original reentrant wave ("F") collide and mutually annihilate.

Another mechanism occurs when the stimulus is applied a little later at the same location (Fig. 4B). Now "A" is able to propagate, but because the DI ahead of it is very short, the APD becomes very short (a phenomenon called restitution[7,11]). After one rotation around the ring, "A" encounters tissue which has very long DI due to the previous short APD. Hence, the

APD now becomes longer — indeed, it becomes sufficiently long that "A" runs into its own refractory back after a second rotation on the ring. Hence, there is a type of alternans amplification,[12] which causes the termination.

While the alternans amplitude is large around $x = 4.5$ cm, the spatial gradients in APD and DI are small. However, there is a large VW (50 ms) when the stimulus is applied after the short action potential at this location. In this case (Fig. 4C), the "A"-wave propagates about halfway around the ring, where it encounters a decreasing DI gradient following "F", causing it to block. In contrast, "R" mutually annihilate with "F" prior to running into the region with decreasing DI, thus ensuring reentry termination. The dynamics are thus qualitatively different from the case of unidirectional block, where "A" fails to propagate from the stimulus site. This example shows that the VW is not only determined by the DI gradient local to the stimulus site, but also depends on more remote gradients.

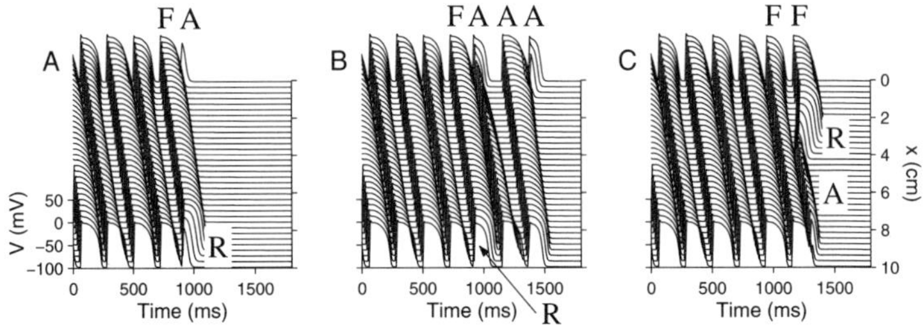

Fig. 4. Termination of reentry during alternans. A: Unidirectional block for $x_s = 0$ cm. B: Alternans amplification where "A" runs into its own tail. $x_s = 0$ cm. C: Conduction block where "A" first propagates, then runs into the back of "F". $x_s = 4.5$ cm.

For some stimulus timing and location combinations, the VW for unidirectional block is decreased compared to the value of 1.6 ms in a longer ring without alternans. This happens when the spatial DI gradient is positive at the stimulus site, so that "A" propagates easily, while "R" blocks. However, in these cases, delivery of a stimulus does not qualitatively alter the dynamics, so that when applying a second stimulus after the next action potential, the VW is similar in size to the single-stimulus case (not shown).

3.3. *Conclusions and outlook*

Depending on the timing and location of the stimulus, the VW for termination may be much increased on the short ring during alternans. Termination may be due to unidirectional block at the stimulus site (Fig. 4A), conduction block with "A" propagating some distance before blocking (Fig. 4C), or alternans-induced block where "A" runs into its own tail (Fig. 4B).

We may not expect the last mechanism to occur in a longer ring since it would require a very long APD. However, rapid pacing in a longer ring will induce APD and DI gradients and we are currently investigating the extent to which these gradients favor termination. In addition, we are examining if pacing with alternating stimulus timing may augment such gradients and increase the VW. While these studies may show the short ring to be a good model of rapid pacing, understanding the mechanisms of termination of alternating reentry is worthwhile in itself since it occurs in the heart in cases of short pathways around obstacles such as valves or scars.

In summary, our findings suggest that current antitachycardia pacing modalities, which use fixed stimulus intervals without considering action potential dynamics, might be improved by protocols that tailor stimulus timing to the dynamics.

Acknowledgments

This work was supported by NSF (PHY-0513389), NIH (R01HL073644), and the Kenny Gordon Foundation (to DJC).

References

1. L. Glass and M. E. Josephson, *Physical Review Letters* **75**, 2059 (1995).
2. T. Nomura and L. Glass, *Physical Review E* **53**, 6353 (1996).
3. M. Courtemanche, L. Glass and J. P. Keener, *Physical Review Letters* **70**, 2182 (1993).
4. R. R. Aliev and A. V. Panfilov, *Chaos, Solitons, & Fractals* **7**, 293 (1996).
5. T. Krogh-Madsen and D. J. Christini, *Physical Review E* **77**, 011916 (2008).
6. Y. Shiferaw, D. Sato and A. Karma, *Physical Review E* **71**, 021903 (2005).
7. M. R. Guevara, G. Ward, A. Shrier and L. Glass, Electrical alternans and period-doubling bifurcations, in *IEEE Computers in Cardiology*, (IEEE Computer Society Press, Los Alamitos, California, 1984) pp. 167–170.
8. L. H. Frame and M. B. Simson, *Circulation* **78**, 1277 (1988).
9. Z. Qu, A. Garfinkel, P.-S. Chen and J. N. Weiss, *Circulation* **102**, 1664 (2000).
10. M. A. Watanabe, F. H. Fenton, S. J. Evans, H. M. Hastings and A. Karma, *Journal of Cardiovascular Electrophysiology* **12**, 196 (2001).
11. M. R. Franz, C. D. Swerdlow, L. B. Liem and J. Schaefer, *Journal of Clinical Investigation* **82**, 972 (1988).
12. P. Comtois and A. Vinet, *Chaos* **12**, 903 (2002).

A PARAMETRIC DERIVATION OF THE SURFACTANT TRANSPORT EQUATION ALONG A DEFORMING FLUID INTERFACE

HUAXIONG HUANG

Department of Mathematics and Statistics, York University, Toronto, Ontario, Canada M3J 1P3

MING-CHIH LAI* and HSIAO-CHIEH TSENG

Department of Applied Mathematics, National Chiao Tung University, 1001, Ta Hsueh Road, Hsinchu 300, Taiwan

A parametric derivation of the surfactant transport equation along a deforming fluid interface is presented in this paper. The derivation is based on the Lagrangian formulation of the interface with a parametric representation. Comparisons with some of the existing derivations are also given.

Keywords: Surfactant transport equation; Surface divergence; Interfacial flow; Parametric representation.

1. Introduction

Surfactant are surface active agents that adhere to the fluid interface and affect the interface surface tension. Surfactant play an important role in many applications in the industries of food, cosmetics, oil, etc. For instance, the daily extraction of ore rely on the subtle effects introduced by the presence of surfactant.[2] In a liquid-liquid system, surfactant allow small droplets to be formed and used as an emulsion. Surfactant also play an important role in water purification and other applications where micro-sized bubbles are generated by lowing the surface tension of the liquid-gas interface. In microsystems with the presence of interfaces, it is extremely important to consider the effect of surfactant since in such case the capillary effect dominates the inertial effect of the fluids.[6]

The basic equation for surfactant transport equation along a deform-

*Corresponding author. E-mail: mclai@math.nctu.edu.tw

ing interface has been derived by Scriven,[4] Aris,[1] and Waxman.[7] All these derivations of the surfactant equation rely heavily on differential geometry. Stone,[5] on the other hand, presented a simple derivation of the time-dependent convective-diffusion equation for surfactant transport along a deforming interface. Stone's derivation leads a form of surfactant mass balance equation which is used later in numerical computations.[9,10] Wong *et al.*[8] derived an alternative form of the surfactant transport equation and provided an interpretation of the equation by Stone.

In this paper, we present a new derivation of the surfactant equation. Our derivation is in the same spirit as Stone's but more detailed. For the immersed interface, we use a global Cartesian position vector and a parametric representation. As a result, the meaning of the time derivative and surface divergence in Stone's equation become clearer. Since an explicit expression for the surface divergence is given, our formulation can be readily incorporated into a front-tracking solver or other interface tracking method for numerical computations, e.g., the immersed boundary method.

2. Surfactant transport equation

Consider a two-dimensional interfacial element $\Sigma(t)$ that is immersed in a three-dimensional incompressible fluid domain Ω. The interface is deformable and moves with the fluid. Following Stone,[5] we assume that the surfactant remains on the interfacial element and does not transport (diffuse) from or to the surrounding bulk fluids, and the total amount on the element is conserved. That is, let Γ denote the mass of the surfactant per unit area, we have

$$\frac{d}{dt}\int_{\Sigma(t)} \Gamma(x,y,z,t)\, dS = 0, \tag{1}$$

where dS is the surface area element. For simplicity, we have also neglected diffusion along the interface. We use two independent parameters (α,β) to label a fixed material point of the initial reference configuration ($\Sigma(0) := \{\boldsymbol{X}_0(\alpha,\beta)|(\alpha,\beta)\in S_0\}$, S_0 is a fixed domain) and the parametric form of the interfacial element at time t is given by $\Sigma(t) := \{\boldsymbol{X}(\alpha,\beta,t)|(\alpha,\beta)\in S_0\}$. In other words, we have used a Lagrangian description of the time evolution of the deformable interface and the following derivation of the surfactant transport equation is based on this parametric form of the interface.

We assume the deforming interface is smooth so that the two independent unit tangent vectors (denoted by $\boldsymbol{\tau}_1$ and $\boldsymbol{\tau}_2$) and its corresponding unit normal vector (denoted by $\boldsymbol{n}$) on the surface can be explicitly expressed by

$$\boldsymbol{\tau}_1 = \frac{\frac{\partial \boldsymbol{X}}{\partial \alpha}}{\left|\frac{\partial \boldsymbol{X}}{\partial \alpha}\right|}, \qquad \boldsymbol{\tau}_2 = \frac{\frac{\partial \boldsymbol{X}}{\partial \beta}}{\left|\frac{\partial \boldsymbol{X}}{\partial \beta}\right|}, \qquad \boldsymbol{n} = \frac{\boldsymbol{\tau}_1 \times \boldsymbol{\tau}_2}{|\boldsymbol{\tau}_1 \times \boldsymbol{\tau}_2|} = \frac{\frac{\partial \boldsymbol{X}}{\partial \alpha} \times \frac{\partial \boldsymbol{X}}{\partial \beta}}{\left|\frac{\partial \boldsymbol{X}}{\partial \alpha} \times \frac{\partial \boldsymbol{X}}{\partial \beta}\right|}. \tag{2}$$

Since the interface is immersed in a three-dimensional incompressible fluid and moves with the local flow velocity, we have

$$\frac{\partial \boldsymbol{X}(\alpha,\beta,t)}{\partial t} = \boldsymbol{u}(\boldsymbol{X}(\alpha,\beta,t),t), \qquad \boldsymbol{X}(\alpha,\beta,0) = \boldsymbol{X}_0(\alpha,\beta), \tag{3}$$

where the fluid velocity $\boldsymbol{u}$ is defined in the fluid domain Ω and satisfies $\nabla \cdot \boldsymbol{u} = 0$. Before we proceed, let us prove the following lemma.

Lemma 2.1. *The material time derivative of the surface element is given by*

$$\frac{d}{dt}\left|\frac{\partial \boldsymbol{X}}{\partial \alpha} \times \frac{\partial \boldsymbol{X}}{\partial \beta}\right| = (\nabla_s \cdot \boldsymbol{u})\left|\frac{\partial \boldsymbol{X}}{\partial \alpha} \times \frac{\partial \boldsymbol{X}}{\partial \beta}\right| \tag{4}$$

where the surface divergent is defined as[a]

$$\nabla_s \cdot \boldsymbol{u} = \left(\frac{\partial \boldsymbol{u}}{\partial \boldsymbol{\tau}_1} \cdot \boldsymbol{b}_2 + \frac{\partial \boldsymbol{u}}{\partial \boldsymbol{\tau}_2} \cdot \boldsymbol{b}_1\right)\left|\frac{\partial \boldsymbol{X}}{\partial \alpha}\right|\left|\frac{\partial \boldsymbol{X}}{\partial \beta}\right| \Big/ \left|\frac{\partial \boldsymbol{X}}{\partial \alpha} \times \frac{\partial \boldsymbol{X}}{\partial \beta}\right|. \tag{5}$$

Here $\boldsymbol{b}_1 = \boldsymbol{n} \times \boldsymbol{\tau}_1$ *and* $\boldsymbol{b}_2 = \boldsymbol{\tau}_2 \times \boldsymbol{n}$ *are the tangential unit vectors normal to* $\boldsymbol{\tau}_1$ *and* $\boldsymbol{\tau}_2$*, respectively, as illustrated in Figure 1.*

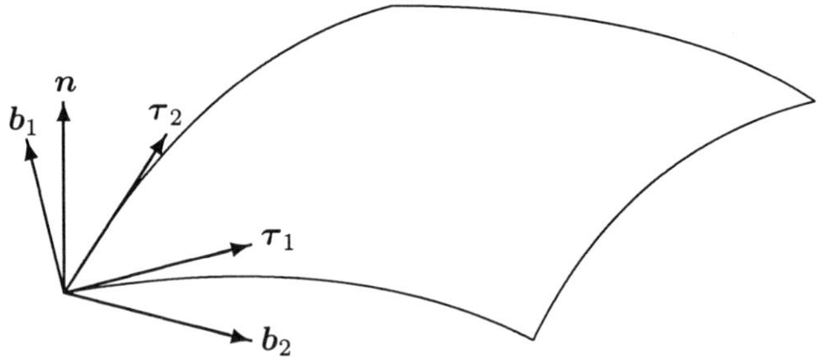

Fig. 1. Illustration of tangential and normal vectors on the surface and their relationships: $\boldsymbol{b}_1 = \boldsymbol{n} \times \boldsymbol{\tau}_1$, $\boldsymbol{b}_2 = \boldsymbol{\tau}_2 \times \boldsymbol{n}$.

[a] A comparison between Eq. (5) and the surface divergence used in the literature[5] is given in Section 3.

Proof.

Firstly, we review some vector identities that will be used in the following. Let us denote $\boldsymbol{a}, \boldsymbol{b}, \boldsymbol{c}$ and $\boldsymbol{d}$ all time-dependent vectors in R^3. The identity

$$(\boldsymbol{a} \times \boldsymbol{b}) \cdot (\boldsymbol{c} \times \boldsymbol{d}) = (\boldsymbol{a} \cdot \boldsymbol{c})(\boldsymbol{b} \cdot \boldsymbol{d}) - (\boldsymbol{a} \cdot \boldsymbol{d})(\boldsymbol{b} \cdot \boldsymbol{c}) \tag{6}$$

can be easily checked and found in.[3] From $\boldsymbol{n} \cdot \boldsymbol{n} = 1$, we have

$$\frac{d\boldsymbol{n}}{dt} \cdot \boldsymbol{n} = 0. \tag{7}$$

We also note that

$$\frac{\partial \boldsymbol{X}}{\partial \alpha} \times \frac{\partial \boldsymbol{X}}{\partial \beta} = \left| \frac{\partial \boldsymbol{X}}{\partial \alpha} \times \frac{\partial \boldsymbol{X}}{\partial \beta} \right| \boldsymbol{n}. \tag{8}$$

We start with the left-hand-side (LHS) of Eq. (4) by using Eqs. (7) and (8) to obtain

$$\begin{aligned}
\frac{d}{dt}\left| \frac{\partial \boldsymbol{X}}{\partial \alpha} \times \frac{\partial \boldsymbol{X}}{\partial \beta} \right| &= \boldsymbol{n} \cdot \frac{d}{dt}\left(\frac{\partial \boldsymbol{X}}{\partial \alpha} \times \frac{\partial \boldsymbol{X}}{\partial \beta} \right) \\
&= \boldsymbol{n} \cdot \left(\frac{\partial^2 \boldsymbol{X}}{\partial t \partial \alpha} \times \frac{\partial \boldsymbol{X}}{\partial \beta} \right) + \boldsymbol{n} \cdot \left(\frac{\partial \boldsymbol{X}}{\partial \alpha} \times \frac{\partial^2 \boldsymbol{X}}{\partial t \partial \beta} \right) \\
&= \boldsymbol{n} \cdot \left(\frac{\partial \boldsymbol{u}}{\partial \alpha} \times \frac{\partial \boldsymbol{X}}{\partial \beta} \right) + \boldsymbol{n} \cdot \left(\frac{\partial \boldsymbol{X}}{\partial \alpha} \times \frac{\partial \boldsymbol{u}}{\partial \beta} \right).
\end{aligned}$$

Substituting $\boldsymbol{n} = \boldsymbol{\tau}_1 \times \boldsymbol{\tau}_2 / |\boldsymbol{\tau}_1 \times \boldsymbol{\tau}_2|$ into the above expression and applying vector identity (6), we obtain

$$\frac{d}{dt}\left| \frac{\partial \boldsymbol{X}}{\partial \alpha} \times \frac{\partial \boldsymbol{X}}{\partial \beta} \right| = G \Big/ \left| \frac{\partial \boldsymbol{X}}{\partial \alpha} \times \frac{\partial \boldsymbol{X}}{\partial \beta} \right| \tag{9}$$

where

$$\begin{aligned}
G = \Bigg[&\frac{\partial \boldsymbol{X}}{\partial \alpha} \cdot \frac{\partial \boldsymbol{u}}{\partial \alpha} \left| \frac{\partial \boldsymbol{X}}{\partial \beta} \right|^2 + \frac{\partial \boldsymbol{X}}{\partial \beta} \cdot \frac{\partial \boldsymbol{u}}{\partial \beta} \left| \frac{\partial \boldsymbol{X}}{\partial \alpha} \right|^2 \\
&- \frac{\partial \boldsymbol{X}}{\partial \alpha} \cdot \frac{\partial \boldsymbol{X}}{\partial \beta} \left(\frac{\partial \boldsymbol{X}}{\partial \beta} \cdot \frac{\partial \boldsymbol{u}}{\partial \alpha} + \frac{\partial \boldsymbol{X}}{\partial \alpha} \cdot \frac{\partial \boldsymbol{u}}{\partial \beta} \right) \Bigg].
\end{aligned}$$

To compute the right-hand-side (RHS) of Eq. (4), it is straightforward to verify that

$$\begin{aligned}
\boldsymbol{b}_1 &= \left(\frac{\partial \boldsymbol{X}}{\partial \alpha} \times \frac{\partial \boldsymbol{X}}{\partial \beta}\right) \times \frac{\partial \boldsymbol{X}}{\partial \alpha} \Big/ \left(\left|\frac{\partial \boldsymbol{X}}{\partial \alpha}\right| \left|\frac{\partial \boldsymbol{X}}{\partial \alpha} \times \frac{\partial \boldsymbol{X}}{\partial \beta}\right|\right) \\
&= \left[\left|\frac{\partial \boldsymbol{X}}{\partial \alpha}\right|^2 \frac{\partial \boldsymbol{X}}{\partial \beta} - \left(\frac{\partial \boldsymbol{X}}{\partial \alpha} \cdot \frac{\partial \boldsymbol{X}}{\partial \beta}\right) \frac{\partial \boldsymbol{X}}{\partial \alpha}\right] \Big/ \left(\left|\frac{\partial \boldsymbol{X}}{\partial \alpha}\right| \left|\frac{\partial \boldsymbol{X}}{\partial \alpha} \times \frac{\partial \boldsymbol{X}}{\partial \beta}\right|\right), \\
\boldsymbol{b}_2 &= \frac{\partial \boldsymbol{X}}{\partial \beta} \times \left(\frac{\partial \boldsymbol{X}}{\partial \alpha} \times \frac{\partial \boldsymbol{X}}{\partial \beta}\right) \Big/ \left(\left|\frac{\partial \boldsymbol{X}}{\partial \beta}\right| \left|\frac{\partial \boldsymbol{X}}{\partial \alpha} \times \frac{\partial \boldsymbol{X}}{\partial \beta}\right|\right) \\
&= \left[\left|\frac{\partial \boldsymbol{X}}{\partial \beta}\right|^2 \frac{\partial \boldsymbol{X}}{\partial \alpha} - \left(\frac{\partial \boldsymbol{X}}{\partial \alpha} \cdot \frac{\partial \boldsymbol{X}}{\partial \beta}\right) \frac{\partial \boldsymbol{X}}{\partial \beta}\right] \Big/ \left(\left|\frac{\partial \boldsymbol{X}}{\partial \beta}\right| \left|\frac{\partial \boldsymbol{X}}{\partial \alpha} \times \frac{\partial \boldsymbol{X}}{\partial \beta}\right|\right).
\end{aligned}$$

Using the definition of the surface divergence of Eq. (5), we find that

$$\nabla_s \cdot \boldsymbol{u} = G \Big/ \left|\frac{\partial \boldsymbol{X}}{\partial \alpha} \times \frac{\partial \boldsymbol{X}}{\partial \beta}\right|^2. \tag{10}$$

Combining Eqs. (9) and (10) completes the proof. □

To derive the governing equation for the surfactant concentration Γ, we begin by applying the law of mass conservation, which yields

$$\begin{aligned}
0 &= \frac{d}{dt} \int_{\Sigma(t)} \Gamma(x, y, z, t)\, dS \\
&= \frac{d}{dt} \int_{\Sigma(0)} \Gamma(\boldsymbol{X}(\alpha, \beta, t), t) \left|\frac{\partial \boldsymbol{X}}{\partial \alpha} \times \frac{\partial \boldsymbol{X}}{\partial \beta}\right| d\alpha\, d\beta \\
&= \int_{\Sigma(0)} \frac{D\Gamma}{Dt} \left|\frac{\partial \boldsymbol{X}}{\partial \alpha} \times \frac{\partial \boldsymbol{X}}{\partial \beta}\right| d\alpha\, d\beta + \int_{\Sigma(0)} \Gamma \frac{d}{dt} \left|\frac{\partial \boldsymbol{X}}{\partial \alpha} \times \frac{\partial \boldsymbol{X}}{\partial \beta}\right| d\alpha\, d\beta,
\end{aligned}$$

where the material derivative is defined as usual $\frac{D\Gamma}{Dt} = \frac{\partial \Gamma}{\partial t}\big|_{\boldsymbol{X}} + \boldsymbol{u} \cdot \nabla \Gamma$, and the subscript $\boldsymbol{X}$ denotes that the derivative is taken with respect to time while $\boldsymbol{X}$ is fixed. In order to simplify the notation, we drop the subscript in the rest of the paper. Using the previous lemma and combining these two integrands, we have

$$\begin{aligned}
0 &= \int_{\Sigma(0)} \left(\frac{\partial \Gamma}{\partial t} + \boldsymbol{u} \cdot \nabla \Gamma + \Gamma(\nabla_s \cdot \boldsymbol{u})\right) \left|\frac{\partial \boldsymbol{X}}{\partial \alpha} \times \frac{\partial \boldsymbol{X}}{\partial \beta}\right| d\alpha\, d\beta, \\
&= \int_{\Sigma(t)} \left(\frac{\partial \Gamma}{\partial t} + \boldsymbol{u} \cdot \nabla \Gamma + \Gamma(\nabla_s \cdot \boldsymbol{u})\right) dS.
\end{aligned}$$

Since the material element is arbitrary, we have derived the equation for the surfactant concentration

$$\frac{\partial \Gamma}{\partial t} + \boldsymbol{u} \cdot \nabla \Gamma + \Gamma \nabla_s \cdot \boldsymbol{u} = 0. \tag{11}$$

Using the fact that the surfactant only transport along the interface and does not transport into the bulk fluid ($\frac{\partial \Gamma}{\partial \boldsymbol{n}} = 0$), we can rewrite the above equation into

$$\frac{\partial \Gamma}{\partial t} + \boldsymbol{u} \cdot \nabla_s \Gamma + \Gamma \, \nabla_s \cdot \boldsymbol{u} = 0. \tag{12}$$

One can conclude that, the physical contributions to the surfactant distribution along the interface come from two parts; namely, due to fluid advection (the second term) and the surface stretching (the third term).

Let us decompose the velocity $\boldsymbol{u}$ into its tangential component along the interface $\boldsymbol{u}_s$, and normal component to the interface $(\boldsymbol{u} \cdot \boldsymbol{n})\boldsymbol{n}$, then Eq. (12) can be expressed as

$$\frac{\partial \Gamma}{\partial t} + \nabla_s \cdot (\Gamma \boldsymbol{u}_s) + \Gamma(\boldsymbol{u} \cdot \boldsymbol{n})(\nabla_s \cdot \boldsymbol{n}) = 0. \tag{13}$$

This leads to the same surfactant transport equation as in.[5] Thus we have established that the time derivative in Stone's derivation is identical to the one used here, i.e., with $\boldsymbol{X}$ fixed.

In,[8] Wong et. al. argued that in Stone's derivation, the time derivative has the meaning that the derivative follows the fixed point which moves normal to the interface rather than the usual meaning of the partial derivative which keeps the surface coordinates (α, β) fixed. Therefore, they derived another surfactant transport equation in which the time derivative is taken by fixing the surface coordinates. The difference between those two derivations can be easily seen by defining $\Gamma(\boldsymbol{X}(\alpha, \beta, t), t) = \hat{\Gamma}(\alpha, \beta, t)$. Going back to the material derivative, we immediately obtain that

$$\frac{D\Gamma}{Dt} + \Gamma \, \nabla_s \cdot \boldsymbol{u} = \frac{\partial \hat{\Gamma}}{\partial t} + \hat{\Gamma} \, \nabla_s \cdot \boldsymbol{u} = 0. \tag{14}$$

It is important to note that the second equation above is exactly the one derived by Wong *et al.* in[8] without the diffusion term.

3. Surface divergence

In this section, we show that our definition of surface divergence is consistent to the one used in the literature[5]

$$\nabla_s \cdot \boldsymbol{u} = (I - \mathbf{n} \otimes \mathbf{n}) \nabla \cdot \boldsymbol{u} = \nabla \cdot \boldsymbol{u} - \boldsymbol{n} \cdot \nabla \boldsymbol{u} \cdot \boldsymbol{n}. \tag{15}$$

In order to simplify the notation, we use z^k as the parameters for the surface with $z^1 = \alpha$ and $z^2 = \beta$. The third parameter z^3 is used for the parameter along the normal direction. We use x^i as the Cartesian coordinates and $\boldsymbol{e}^i$

as the corresponding unit vectors for $i = 1, 2, 3$. Therefore, the position and velocity vectors can be expressed as

$$\boldsymbol{X} = x^i \boldsymbol{e}^i, \quad \boldsymbol{u} = u^i \boldsymbol{e}^i$$

where the repeated indices indicate the summation. To deal with non-Cartesian coordinates one often uses the co-variant and contra-variant basis vectors as follows

$$\boldsymbol{g}_k = \frac{\partial x^i}{\partial z^k}\boldsymbol{e}^i, \quad \boldsymbol{g}^k = \frac{\partial z^k}{\partial x^i}\boldsymbol{e}^i, \quad k = 1, 2, 3 \tag{16}$$

with the orthogonal property

$$\boldsymbol{g}^i \cdot \boldsymbol{g}_j = \delta^i_j \tag{17}$$

where δ^i_j is the Kronnecker delta symbol. Comparing with our earlier notation, we have

$$\boldsymbol{\tau}_1 = \frac{\boldsymbol{g}_1}{|\boldsymbol{g}_1|}, \ \boldsymbol{\tau}_2 = \frac{\boldsymbol{g}_2}{|\boldsymbol{g}_2|}, \ \boldsymbol{b}_1 = \frac{\boldsymbol{g}^2}{|\boldsymbol{g}^2|}, \ \boldsymbol{b}_2 = \frac{\boldsymbol{g}^1}{|\boldsymbol{g}^1|}, \ \boldsymbol{n} = \frac{\boldsymbol{g}_3}{|\boldsymbol{g}_3|} = \frac{\boldsymbol{g}^3}{|\boldsymbol{g}^3|}.$$

Using Eq. (16) and the fact that $\boldsymbol{e}^i$ are constant unit vectors, we have

$$\nabla \cdot \boldsymbol{u} = \frac{\partial(\boldsymbol{e}^i \cdot \boldsymbol{u})}{\partial x^i} = \frac{\partial(\boldsymbol{e}^i \cdot \boldsymbol{u})}{\partial z^k}\frac{\partial z^k}{\partial x^i} = \boldsymbol{e}^i \cdot \frac{\partial \boldsymbol{u}}{\partial z^k}\frac{\partial z^k}{\partial x^i} = \boldsymbol{g}^k \cdot \frac{\partial \boldsymbol{u}}{\partial z^k}.$$

Similarly, we have

$$\nabla \boldsymbol{u} = \frac{\partial \boldsymbol{u}}{\partial x^i}\boldsymbol{e}^i = \frac{\partial \boldsymbol{u}}{\partial z^k}\boldsymbol{g}^k$$

and

$$\boldsymbol{n} \cdot \nabla \boldsymbol{u} \cdot \boldsymbol{n} = \boldsymbol{n} \cdot \frac{\partial \boldsymbol{u}}{\partial z^k}\boldsymbol{g}^k \cdot \boldsymbol{n} = \boldsymbol{n} \cdot \frac{\partial \boldsymbol{u}}{\partial z^3}\left|\boldsymbol{g}^3\right| = \boldsymbol{g}^3 \cdot \frac{\partial \boldsymbol{u}}{\partial z^3}.$$

Therefore,

$$\begin{aligned}
\nabla \cdot \boldsymbol{u} - \boldsymbol{n} \cdot \nabla \boldsymbol{u} \cdot \boldsymbol{n} &= \boldsymbol{g}^1 \cdot \frac{\partial \boldsymbol{u}}{\partial z^1} + \boldsymbol{g}^2 \cdot \frac{\partial \boldsymbol{u}}{\partial z^2} \\
&= \left|\boldsymbol{g}^1\right| \left|\boldsymbol{g}_1\right| \boldsymbol{b}_2 \cdot \frac{\partial \boldsymbol{u}}{\partial \boldsymbol{\tau}_1} + \left|\boldsymbol{g}^2\right| \left|\boldsymbol{g}_2\right| \boldsymbol{b}_1 \cdot \frac{\partial \boldsymbol{u}}{\partial \boldsymbol{\tau}_2}.
\end{aligned} \tag{18}$$

From Eq. (17), we have

$$\left|\boldsymbol{g}^1\right| \left|\boldsymbol{g}_1\right| \boldsymbol{\tau}_1 \cdot \boldsymbol{b}_2 = \left|\boldsymbol{g}^2\right| \left|\boldsymbol{g}_2\right| \boldsymbol{\tau}_2 \cdot \boldsymbol{b}_1 = 1.$$

Using

$$\boldsymbol{\tau}_1 \cdot \boldsymbol{b}_2 = \boldsymbol{\tau}_1 \cdot (\boldsymbol{\tau}_2 \times \boldsymbol{n}) = \boldsymbol{n} \cdot (\boldsymbol{\tau}_1 \times \boldsymbol{\tau}_2) = |\boldsymbol{\tau}_1 \times \boldsymbol{\tau}_2|$$

and

$$\boldsymbol{\tau}_2 \cdot \boldsymbol{b}_1 = \boldsymbol{\tau}_2 \cdot (\boldsymbol{n} \times \boldsymbol{\tau}_1) = \boldsymbol{n} \cdot (\boldsymbol{\tau}_1 \times \boldsymbol{\tau}_2) = |\boldsymbol{\tau}_1 \times \boldsymbol{\tau}_2|,$$

we obtain

$$|\boldsymbol{g}^1|\,|\boldsymbol{g}_1| = |\boldsymbol{g}^2|\,|\boldsymbol{g}_2| = 1/\,|\boldsymbol{\tau}_1 \times \boldsymbol{\tau}_2| = \left|\frac{\partial \boldsymbol{X}}{\partial \alpha}\right|\left|\frac{\partial \boldsymbol{X}}{\partial \beta}\right| \Big/ \left|\frac{\partial \boldsymbol{X}}{\partial \alpha} \times \frac{\partial \boldsymbol{X}}{\partial \beta}\right|. \quad (19)$$

Combining Eq. (19) and Eq. (18) shows that the surface divergence given by Eq. (15) is consistent with our definition (5).

Acknowledgments

M.-C. Lai is supported in part by National Science Council of Taiwan under research grant NSC-95-2115-M-009-010-MY2 and H. Huang is supported by grants from the Natural Science and Engineering Research Council (NSERC) of Canada and the Mathematics of Information Technology and Complex Systems (MITACS) of Canada.

References

1. R. Aris, *Vectors, Tensors, and the Basic Equations of Fluid Mechanics*, Prentice-Hall, Englewood Cliffs, NJ, 1962.
2. P. G. De Gennes, F. Brochard, D. Quere, *Gouttes, Bulles, Perles Ondes*, Edition Berlin, 2002.
3. A. Jeffrey, Handbook of Mathematical Formulas and Integrals, 3rd Ed., Elsvier, (2004) 361.
4. L. E. Scriven, Dynamics of a fluid interface, *Chem. Eng. Sci.*, 12, (1960), 98.
5. H. A. Stone, A simple derivation of the time-dependent convective-diffusion equation for surfactant transport along a deforming interface, *Phys. Fluids A*, 2(1), (1990) 111–112.
6. P. Tabeling, *Introduction to Microfluidics*, Oxford University Press, 2005.
7. A. M. Waxman, Dynamics of a couple-stress fluid membrane, *Stud. Appl. Math.* 70, (1984) 63.
8. H. Wong, D. Rumschitzki and C. Maldarelli, On the surfactant mass balance at a deforming fluid interface, *Phys. Fluids*, 8(11), (1996) 3203–3204.
9. J.-J. Xu, Z. Li, J. S. Lowengrub and H.-K. Zhao, A level-set method for interfacial flows with surfactant, *J. Comput. Phys.*, vol 212, (2006) 590–616.
10. J.-J. Xu and H.-K. Zhao, An Eulerian formulation for solving partial differential equations along a moving interface, *J. Sci. Comput.*, vol 19, (2003) 573–593.

ANALYSIS AND ESTIMATION OF THE VARIANCE OF CROSS-VALIDATION ESTIMATORS OF THE GENERALIZATION ERROR: A SHORT REVIEW

MARIANTHI MARKATOU* and ROSITSA DIMOVA**

Department of Biostatistics, Columbia University,
New York, NY 10032, USA
**E-mail: mm168@columbia.edu, **E-mail: rbd2107@columbia.edu*

ANSHU SINHA+

Department of Biomedical Informatics, Columbia University
New York, NY 10032, USA
+E-mail: anshu.sinha@dbmi.columbia.edu

We briefly review research on the estimation of variance of cross validation estimators of the generalization error of computer algorithms. A general methodology for analyzing and estimating the variance of cross validation estimators of the generalization error is also discussed in some detail.

Keywords: Classification, Cross-validation, Estimation, Generalization error, Regression.

1. Introduction

The problem of estimating the variance of estimators of the generalization error is important because it aids the selection of algorithms with better generalizability properties. In order to compare learning algorithms, statistical tests of significance are often used (Dietterich[1]). To obtain accurate results from these tests, we need to use appropriate methods to accurately estimate the variability of the quantities under comparison. The generalization error is a measure of algorithmic performance, and it is very often used to compare different algorithms, the idea being the smaller the generalization error of an algorithm, the better the algorithm performs.

While a lot has been written on the topic of estimation of the generalization error, much less has been written on the subject of estimating the variance of estimators of the generalization error. This is due to the difficulty of creating a general theory for the estimation of the variance of the estimators of the generalization error. Some of the difficulties encountered are due to the desire to take into account all sources of variability, such as the choice of the training set (Breiman[2]) or initial conditions of a learning algorithm (Kolen & Pollack[3]). Some distribution free bounds on the deviations of cross-validation (CV) are available, but they are specific to locally defined classifiers, such as nearest-neighbors (Kohavi[4]).

The problem of variance estimation of the generalization error is not new. In a series of interesting papers, McLachlan[5,6] addressed the problem of estimation of the variance of the errors of misclassification of the linear discriminant function. He developed a technique for deriving asymptotic expansions of the variances of misclassification errors of Anderson's statistic. Recently, Nadeau & Bengio[7] investigated the theoretical and practical merits of various estimators of the variance of CV estimators of the generalization error, taking into account the variability due to the choice of *training and test sets*. Markatou *et al.*[8] studied one particular variance estimator proposed by Nadeau & Bengio[7] and developed a general approach to this problem. This approach is related to the one developed by McLachlan[5,6] and views the estimation of variance as a problem in approximating the moments of a statistic; in this context, the statistic under consideration is the CV generalization error. Additionally, Bengio & Grandvalet[9] proved that there does not exist an unbiased estimator of the variance of the k-fold cross validation.

This paper is organized as follows: Section 2 introduces notation and various concepts that will be used throughout the paper. Section 3 briefly discusses one particular estimator suggested by Nadeau & Bengio,[7] and then presents the analysis suggested by Markatou *et al.*[8] Section 4 offers a discussion and conclusions.

2. Notation and Definitions

Let the data $x_1, x_2, ..., x_n$ be realizations of the random variables $X_1, X_2, ..., X_n$, collected in the set $Z_1^n = \{X_1, X_2, ..., X_n\}$, called the *data universe.* The random variables are assumed to be independent and identically distributed but no other assumptions are made on their distribution F. Let S represent a subset of size n_1, $n_1 < n$, taken from Z_1^n. Denote by S^c the complement of the set S, with respect to the data universe. In what follows, we define the terminology that we will use.

Definition 2.1. Any subset S of Z_1^n that is used to construct a statistical/algorithmic rule is called a training set. The cardinality of the training set is $n_1 < n$.

Definition 2.2. A data set, collected independently from the training set data, but with the same underlying distribution as that of the training set data which is used to assess the constructed rule, is called the test set data. The cardinality of the test set is $n_2 = n - n_1$.

Definition 2.3. A prediction rule or model that is constructed using the data in a training set S is a measurable function f such that f:S$\rightarrow$ R.

Definition 2.4. Let $L : R^p \times R \rightarrow R$ be a function, let Y be a target variable and $\hat{f}(X)$ a prediction rule. The function $L(Y, \hat{f}(X))$ that measures the error between the target variable Y and $\hat{f}(X)$ is called a loss function.

The most popular form of loss function is the square error loss, defined as $L(Y, \hat{f}(X)) = (Y - \hat{f}(X))^2$. Other typical choices of loss functions include the absolute error loss or the indicator loss that is used in classification.

Definition 2.5 (Hastie *et al.*[10]). *Generalization error or test error is the expected prediction error over an independent test sample; it is given by the quantity $E[L(Y, \hat{f}(X))]$, where Y, X are drawn randomly from their joint distribution and the expectation is taken over everything that is random.*

The generalization error can be estimated either via bootstrapping or via cross-validation. In this paper we will concentrate on the CV estimators of the generalization error. In particular, we offer estimators of the variance for the following two CV schemes.

Non-overlapping test set selection: In this case, if $S_j^c, S_{j'}^c$, $j \neq j'$ are two different test sets, $S_j^c \cap S_{j'}^c = \varnothing$. This case corresponds to k-fold cross-validation, where the data are divided into k non-overlapping blocks of approximately equal size. Each of these blocks is used as a test set, and the remaining data are used to train the algorithm.

Complete Random Selection: In this scheme the training sets, and therefore the test sets, are selected completely at random, and without replacement, from the set of the data universe.

The CV estimator of the generalization error we will study is defined as follows. Let A_j be a random set of n_1 distinct integers, n_1 is fixed. Select $j = 1, 2, ..., J$ of these sets at random, and then from the data universe select the corresponding data elements. Denote by $L(j, i)$ the loss function that measures the difference between the rule trained on the j-th training set S_j and tested on the elements of the corresponding test set. The CV estimator of the generalization error is then defined as:

$$\hat{T}_J = \frac{1}{J}\sum_{j=1}^{J}\frac{1}{n_2}\sum_{i\in S_j^c} L(j,i) = \frac{1}{J}\sum_{j=1}^{J}\hat{\mu}_j,$$

where $\hat{\mu}_j$ is the usual average test set error measured on the test set S_j^c, J is the number of different training (and test) sets used. The next section discusses one particular variance estimator of this CV estimator of the generalization error introduced by Nadeau & Bengio[7] and an approach to variance estimation that unifies both schemes of CV given above, introduced by Markatou *et al.*[8]

3. Variance Estimation of $\hat{\mathbf{T}}_J$

Nadeau & Bengio[7] proposed new estimators of the variance of $\hat{T}_J$, one of which is given as follows. Let

$$S^2_{\hat{\mu}_j} = \frac{1}{J-1}\sum_{j=1}^{J}(\hat{\mu}_j - \hat{T}_J)^2,$$

be the sample variance of $\hat{\mu}_j, j = 1, 2, ..., J$. If $\rho = Corr(\hat{\mu}_j, \hat{\mu}_{j'})$, $j \neq j'$, we obtain that an unbiased estimator of the variance of $\hat{T}_J$, under the assumption that ρ is known, is given by the formula

$$(\frac{1}{J} + \frac{\rho}{1-\rho})S^2_{\hat{\mu}_j}.$$

Nadeau & Bengio[7] observe that the correlation ρ is difficult to estimate; they offer the approximation $\rho = n_2/n$ stating that it will tend to estimate almost unbiasedly the correlation if the prediction rule does not change much when different training sets are chosen. They recommend using J=15, and utilizing the suggested approximation to the correlation their estimator can be written as

$$(\frac{1}{J} + \frac{n_2}{n_1})S^2_{\hat{\mu}_j}.$$

Markatou *et al.*[8] treat $\hat{T}_J$ as a statistic and analyze its variance using the identity

$$Var(\hat{T}_J) = \frac{1}{J^2}\sum_{j=1}^{J} Var(\hat{\mu}_j) + \frac{1}{J^2}\sum_{j\neq j'}\sum Cov(\hat{\mu}_j, \hat{\mu}_{j'}).$$

From this formula, it can be seen that if the terms in it can be approximated, we automatically obtain an estimator of the variance of $\hat{T}_J$. This estimator is not an unbiased estimator; it is an approximately unbiased estimator with bias order $1/n^2$.

To approximate the variance and covariance terms involved, the CV estimator is viewed as a statistic, and the moments of its distribution are approximated. The approximations illustrate the role of training and test sets in the performance of the algorithm and take into account the variability due to different training and test sets.

Markatou *et al.*[8] carried out the analysis suggested above for a variety of cases including prediction of the sample mean, regression and classification. Although this analysis is carried out for differentiable loss functions, non-differentiable, at a finite number of points, loss functions can be treated similarly, by approximation with differentiable loss functions, so there is no restriction from the assumption of smoothness imposed on the loss functions. In particular, Markatou *et al.*[8] proved that in the case of prediction of the sample mean the variance of the estimator $\hat{T}_J$ is a function of the moments of the distribution of the random variables $Y_j = Card(S_j \cap S_{j'})$ and $Y_j^* = Card(S_j^c \cap S_{j'}^c)$, $j \neq j'$. The distribution of these random variables is hypergeometric, linking the problem with the capture-recapture problem in statistics.

Recall that no assumptions on the distribution of the random variables the realizations of which constitute our data were made by Nadeau & Bengio.[7] Markatou *et al.*[8] require the existence of the expectation for derivatives of the loss function L, and existence of expectations of the loss function itself. No specific assumptions on the distribution of the data are needed.

As an example, we present now the estimator of $Var(\hat{T}_J)$, for the problem of prediction of the sample mean, when the loss function is squared error, and under both CV schemes reviewed in the previous section. The estimator of variance under completely random cross-validation is given as

$$V\hat{ar}(\hat{T}_J) = c_1\hat{M}_1 + c_2\hat{M}_2,$$

where

$$\hat{M}_1 = \frac{1}{J}\sum_{j=1}^{J}\frac{1}{n_1}\sum_{i\in S_j}(X_i - \bar{X}_{S_j})^4,$$

$$\hat{M}_2 = \{\frac{1}{J}\sum_{j=1}^{J}\frac{1}{n_1 - 1}\sum_{i\in S_j}(X_i - \bar{X}_{S_j})^2\}^2.$$

Moreover, the constants are given as $c_1 = (1/n)(1 + (n_1/Jn_2))$, and $c_2 = (1/J)\{[4/n_1n_2] + (J-1)[4n_1^2 - n^2 - nn_1 + n]/n_1^2n^2\}$.

The estimator of variance under non-overlapping test set selection (k-fold cross-validation, J=k), and quadratic error loss is given as

$$\frac{1}{n}\{\sum_{j=1}^{k}\frac{1}{n_1}\sum_{i\in S_j}(X_i - \bar{X}_{S_j})^4\} + \frac{1}{n}(\frac{3k}{n(k-1)} - 1)\{\sum_{j=1}^{k}\frac{1}{n_1 - 1}\sum_{i\in S_j}(X_i - \bar{X}_{S_j})^2\}^2\}.$$

These estimators provide more accurate results than the Nadeau & Bengio[7] estimator, as their variance is always smaller than that of the Nadeau & Bengio estimator. Further discussion and details on the estimators for the case of regression and classification can be found in Markatou *et al.*[8] and Tian.[11]

4. Summary and Conclusions

We reviewed briefly the problem of variance estimation of the CV estimators of generalization error. This problem is complicated for a variety of reasons including the various sources of variability that we desire to take into account. In addition, the generalization error and its variance depend on the size of training and test sets and the loss function that is used. The ideas outlined in Markatou *et al.*[8] and in Tian[11] can provide estimators in general settings such as kernel regression and classification.

Acknowledgments

Dr. Markatou and R. Dimova are supported by NSF DMS-0504957 and A. Sinha is supported by the NLM Informatics Research Training Program (5 T15 LM007079-15). The authors would like to thank Dr. E. Michalopoulou and the organizers of "Frontiers in Applied and Computational Mathematics 2008" for the invitation to present this work.

References

1. T. G. Dietterich, *Neural Computation* **10**, 1895 (1998).
2. L. Breiman, *Annals of Statistics* **24**, 2350 (1996).
3. J. Kolen and J. Pollack, *Back propagation is sensitive to initial conditions.*, in *Advances in Neural Information Processing Systems*, (nMorgan Kauffmann, San Francisco, CA, 1991), San Francisco, CA.
4. R. Kohavi, A study of cross-validation and bootstrap for accuracy estimation and model selection., in *Proceedings of the Fourteenth International Joint Conference on Artificial Intelligence*, 1995.
5. G. J. McLachlan, *Australian Journal of Statistics* **14**, 68 (1972).
6. G. J. McLachlan, *Australian Journal of Statistics* **15**, 210 (1974).
7. C. Nadeau and Y. Bengio, *Machine Learning* **52**, 239 (2003).
8. T. H. B. S. Markatou, M. and G. Hripcsak, *Journal of Machine Learning Research* **6**, 1127 (2005).
9. Y. Bengio and Y. Grandvalet, *Journal of Machine Learning Research* **5**, 1089 (2004).
10. R. T. T. Hastie and J. Friedman, *The Elements of Statistical Learning: Data Mining, Inference and Prediction.*Springer Series in Statistics, Springer Series in Statistics (Springer-Verlag, New York, 2001).
11. H. Tian, Variance estimation of the cross validation estimator of the generalization error., PhD thesis, Department of Biostatistics, Columbia University, (NY, USA, 2006).

NEGATIVE PHASE AND LEADER SWITCHING IN NON-WEAKLY COUPLED TWO-CELL INHIBITORY NETWORKS *

VICTOR MATVEEV[‡] and MYONGKEUN OH

Department of Mathematical Sciences, New Jersey Institute of Technology
University Heights, Newark, NJ 07102-1982, U.S.A.
[‡]*E-mail: matveev@njit.edu*

We examine the dynamics of a non-weakly coupled inhibitory network of two identical Morris-Lecar model neurons with type-I excitability, which was recently shown to exhibit stable alternating-order activity, whereby the spiking order of the two cells changes in each cycle of the oscillation. We provide an intuitive geometric description of such *leader switching* and demonstrate that the concept of negative phase allows to analyze the existence and stability of such alternating-order dynamics.

Keywords: non-weakly coupled oscillators, inhibitory network, leader switching, synchronization

1. Introduction

Understanding the dynamics and synchronization in inhibitory networks is a question of fundamental importance in neuroscience, since such networks play a crucial role in rhythmic activity in a variety of neural systems, from invertebrate central pattern generators (CPGs) to the mammalian brain.[2] Invertebrate CPGs in particular often contain simple two-cell inhibitory sub-networks that control different rhythmic motor behaviors. Therefore, characterizing the dynamics of such networks is relevant for a better understanding of the rhythmogenesis in biological inhibitory circuits. While recent work has revealed the synchronizing role of the inhibitory synaptic interaction,[2] it is known that non-weak coupling can destabilize phase-locked dynamics. In particular, Maran and Canavier[1] recently examined an inhibitory network of Wang-Buzsáki model neurons with type-I excitability,[3] and demonstrated that this network exhibits 2:2 mode-locked states

*Supported by the National Science Foundation grant DMS-0417416.

whereby the firing order of the coupled cells changes in each cycle of oscillation (see Fig. 1). We find that such behavior (termed "leap-frog" spiking or "leader switching"[4]) is a more general property of a subclass of non-weakly coupled type-I oscillators characterized by slow dynamics of membrane potential upon hyperpolarization, and in particular can be achieved in a network of simpler Morris-Lecar (ML) model neurons. We present an intuitive geometric description of leader switching by examining the phase space trajectory of the two cells, and establish the most fundamental conditions for the existence and stability of such dynamic behavior.

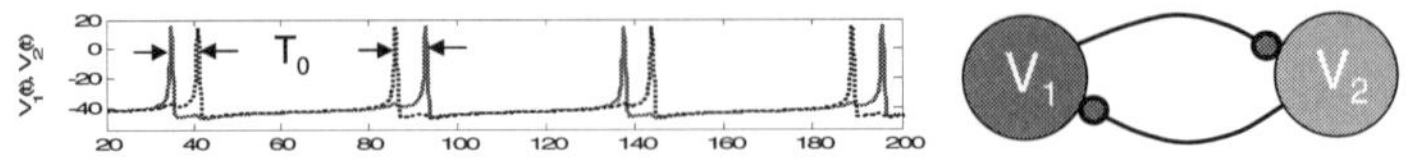

Fig. 1. Time-course of membrane potentials of the two non-weakly coupled ML cells, $V_1(t)$ and $V_2(t)$, showing periodic leader switching ("leap-frog" spiking). Note that the interval between two spikes of the same cell equals the intrinsic (unperturbed) oscillation period, T_0.

2. Phase-plane geometry of alternating-order spiking

Figure 2A shows the phase-space geometry of the ML model neuron (equations and parameters given in Appendix). This model has two dynamical degrees of freedom: the membrane potential, V, and the activation of the outward K^+ current, w. The V-nullcline is cubic-shaped, while the w-nullcline is sigmoidal. In the absence of synaptic coupling, the V-nullcline is in its upper position; the two nullclines intersect at an unstable fixed point, and there exists a stable limit cycle shown in blue, which represents an electric pulse (an *action potential*; see V time-course in Fig. 1). This limit cycle results from the push-pull interaction between the voltage-activated inward Na^+ current and the more slowly activating outward K^+ current (first two terms in the Kirchhoff current balance, Eq. A.1).

The two identical ML cells are coupled by inhibition, which means that a V spike in one cell (the *presynaptic* cell) will introduce a hyperpolarizing current into its partner cell (the *postsynaptic* cell). This has the effect of lowering the V nullcline of the postsynaptic cell (gray curve in Fig. 2A), which in its lowered position intersects the w nullcline at a *stable* fixed point, where the postsynaptic cell will be transiently trapped. This transient suppression (lasting for the duration of the synaptic current) allows the

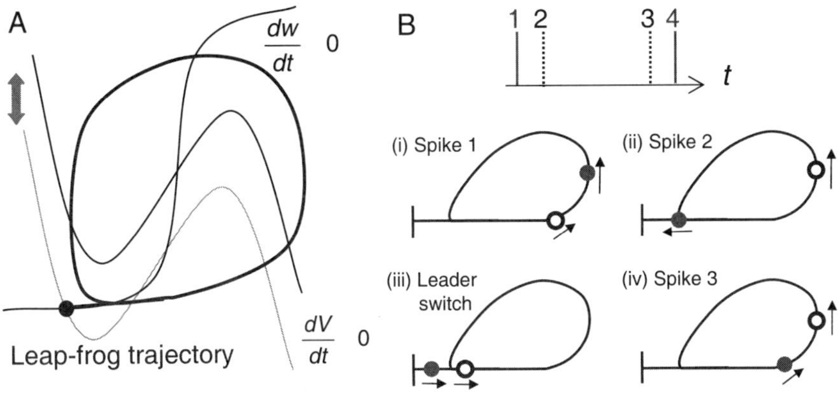

Fig. 2. Phase plane dynamics of each ML cell during alternating-order spiking. (A) Nullclines of the ML model. Synaptic inhibition lowers the V-nullcline of a cell, transiently trapping it at a stable fixed point. Tadpole-shaped blue curve indicates the trajectory of each cell during one cycle of the alternating-order spiking shown in Fig. 1. Note that it overlaps the w-nullcline during the hyperpolarized phase of the oscillation. (B) The sequence of four panels describes the leap-frog spike sequence at the top, with filled red and open blue circles representing the two cells: (i) "red" cell spikes; (ii) "blue" cell spikes, pushing the "red" cell into the subthreshold branch of the trajectory (tadpole tail); (iii) "blue" cell bypasses the "red" cell along the unperturbed limit cycle; (iv) "blue" cell spikes again. The process then repeats itself, with the "red" cell spiking next (color can be viewed from the e-version).

presynaptic cell to advance past the postsynaptic cell along its trajectory, as described by the sequence of panels in Fig. 2B.

Interestingly, the mechanism just described is valid even in the limit of infinitely short synaptic interaction, whereby the synaptic current introduces an instantaneous hyperpolarization into the postsynaptic cell. This is a surprising fact, since such an instantaneous input produces a shift in the potential, but does not lower the V-nullcline, and therefore does not create a meta-stable fixed point where the cell could be transiently trapped. To understand the existence of leap-frog spiking in this *pulse-coupled* case, we examined the *isochrons* of the model, shown in Fig. 3. An isochron is a powerful concept allowing to analyze the response of oscillators to perturbations;[5–7] it represents a set of points that asymptotically approach the same point on the limit cycle. An isochron foliation therefore completely describes the long-term behavior of the system starting at any initial position in the basin of attraction of the limit cycle, in terms of the point on the cycle that this trajectory will eventually approach. Now, if the presynaptic cell spikes while the postsynaptic cell is on the bottom part of the

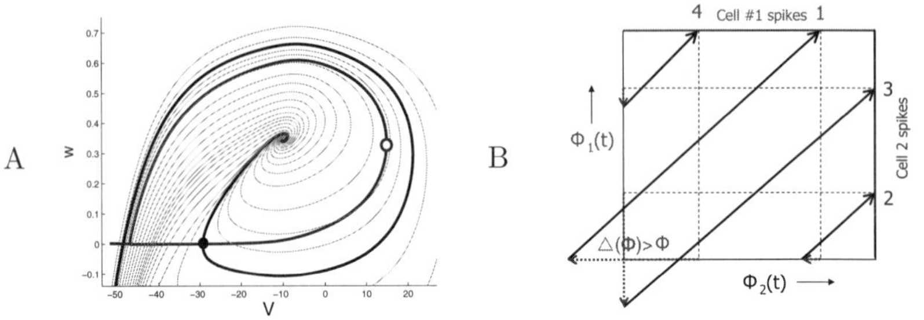

Fig. 3. Leap-frog spiking in a pulse-coupled network. (A) Isochron foliation of the model ML cell. Note the characteristic curling of the isochrons around the limit cycle. (B) Phase-reduced description of the alternating-order spike sequence in top panel of Fig. 2B. Top and right boundaries correspond to spikes of cell 1 (red) and cell 2 (blue), respectively. The trajectory is discontinuous across these boundaries, due to synaptic inhibition received at each boundary crossing (arrows) (color can be viewed from the e-version).

trajectory (where it in fact spends most of the time), inhibition produces a left-ward shift (negative V pulse) onto an isochron that curls around the limit cycle, intersecting it at a position (filled circle) that is retrograde to the position of the presynaptic cell at spike-time (open circle). This means that the inhibition retards the dynamics of the postsynaptic cell sufficiently to allow the presynaptic cell to advance ahead of it, reversing the spiking order of the two cells. Such strong retardation is a result of slow dynamics of V along the portion of the limit cycle overlapping the w-nullcline, which constitutes the slow manifold of the system, due to fast closing of K^+ channels (fast w dynamics) at hyperpolarized potentials.

If the oscillators stay close to their limit cycles, isochrons allow to reduce the n-dimensional state space to a one-dimensional *phase* variable defining position along the limit cycle, since there is a (surjective) mapping from any state-space position onto the corresponding isochron. This 1-D phase variable can be simply defined as the time variable of the limit cycle trajectory, normalized by the unperturbed oscillation period. Such dimensional reduction is in fact the key to analyzing weakly-coupled oscillator dynamics,[5,7] but we also find it useful in our case of non-weak coupling. In Figure 3B, we qualitatively describe leader switching as a trajectory on the surface of a 2-D phase torus formed by the 1-D phase variables of each cell. The top boundary of the torus corresponds to the spike of cell 1, which sends an inhibitory pulse to cell 2, pushing the trajectory leftward as it is resets (wraps around) in the vertical direction. Analogously, the right boundary corresponds to the spike of cell 2, which sends an inhibitory pulse to cell 1, producing a downward trajectory shift. Note that leader switch only occurs if these synaptic deflections (phase resets, $\Delta(\varphi)$) are strong enough to push the trajectory outside of the phase domain, into the region corresponding to *negative* phase values. Such strong phase resets correspond to the isochrons intersecting the limit cycle at a position retrograde to the peak of an action potential, as described above (panel A). Finally, note that in the absence of coupling the trajectory is a straight line of slope 1, given the simple definition of phase as normalized time.

3. Existence and stability of periodic leader switching

Figure 4A "unwraps" the phase-torus in Fig. 3B, plotting the phase time-course for one of the two cells. Phase is reduced (delayed) by amounts Δ in response to each spike of the partner cell. The dependence of phase delay on cell's current phase, $\Delta(\varphi)$, is shown in Fig. 4B; it is known as the *the spike-time response curve* (STRC). It is computed numerically by measuring the lengthening of the cell's period produced by synaptic inputs applied at

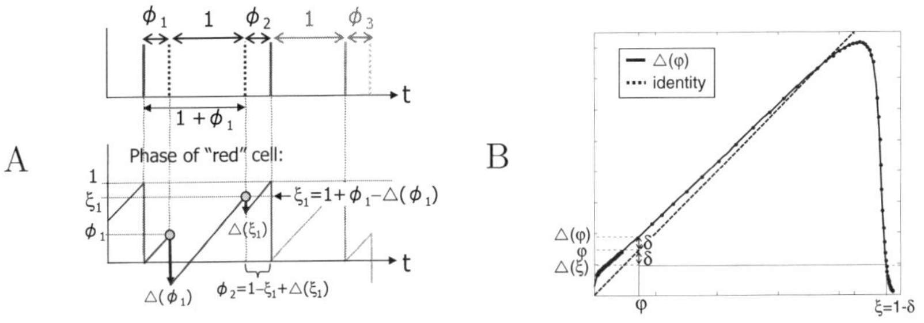

Fig. 4. Phase-map analysis of alternating-order spiking. (A) Spike sequence (top) and the phase time-course of one of the two cells (bottom) during leap-frog spiking. During one oscillation cycle, each cell spikes twice between two spikes of the partner cell. Phase intervals φ_i are inter-spike intervals normalized by the unperturbed period of each oscillator. The phase difference between two dashed spikes equals 1 (the unperturbed period). Phase delays due to each of the two spikes (blue arrows) equal $\Delta(\varphi_1)$ and $\Delta(\xi_1)$, where ξ_1 is the phase of the cell at the time of arrival of the second input, $\xi_1 = 1 + \varphi_1 - \Delta(\varphi_1)$. The second inter-spike interval φ_2 is found by the first-passage time condition $\xi_1 - \Delta(\xi_1) + \varphi_2 = 1$. (B) STRC of each ML cell, $\Delta(\varphi)$, for $\bar{g}_{syn} = 0.2$. Equilibrium inter-spike phase difference ($\varphi = 0.144$) in the alternating-order state satisfies Eq. 3. Note that $\delta = \Delta(\varphi) - \varphi = \varphi - \Delta(\xi)$, where ξ is the phase of the postsynaptic cell at the time of arrival of the second spike, $\xi = 1-\delta$. Here $\delta = 0.0468$, and $\Delta(1-\delta) = 0.095$ (color can be viewed from the e-version).

different times (phases) since the peak of the cell's action potential, which is defined as the zero phase point. The STRC is a powerful tool in analyzing both weakly and non-weakly coupled oscillator dynamics.[4–8] Following Ref. 1, we will now use the STRC to analyze the existence and stability of leap-frog spiking, but will restrict ourselves to the case of a homogeneous network for simplicity. Our derivation is a direct extension of Ref. 8 to the case of negative phase values, previously viewed as problematic.

Alternating-order firing is completely characterized by the inter-spike phase sequence labeled $\{\varphi_1, \varphi_2\}$ in Fig. 4A; we will construct the return map relating these phase intervals. Note that ϕ_1 is the phase of cell 1 (*red* spike and *red* trace) at the arrival time of the first synaptic pulse from the pre-synaptic cell (dashed blue line), where phase is defined as time since preceding spike, normalized by unperturbed oscillation period. The amount of phase delay induced by this input equals $\Delta(\varphi_1)$. For sufficiently strong synaptic inhibition this phase reset satisfies $\Delta(\varphi_1) > \varphi_1$ which delays the first passage time to next spike of cell 1 to a value greater than 1, the intrinsic oscillation period. As a result, cell 2 has a chance to spike again (second dashed line), after a phase interval corresponding to the unperturbed oscillation period, $\Delta\varphi = 1$, since cell 2 receives no input from cell 1 during this period. This second synaptic current arrives when the phase of cell 1 equals $\xi_1 \equiv 1 + \varphi_1 - \Delta(\varphi_1)$, taking into account the delay due to the first spike; therefore, it induces a phase delay equal to $\Delta(1 + \varphi_1 - \Delta(\varphi_1))$. It is only after receiving this second input that cell 1 finally has a chance to spike, after a phase interval defined as φ_2. The total phase delay due to both inputs is thus equal to $\varphi_1 + \varphi_2 = \Delta(\varphi_1) + \Delta(1 + \varphi_1 - \Delta(\varphi_1))$. Therefore, the return map for the phase intervals φ_i is given by

$$\varphi_2 \equiv \Phi(\varphi_1) = \Delta(\varphi_1) + \Delta(1 + \varphi_1 - \Delta(\varphi_1)) - \varphi_1 \tag{1}$$

or, expressed in terms of the phase of the post-synaptic cell at the time of arrival of the second spike, $\xi_1 = 1 + \varphi_1 - \Delta(\varphi_1)$:

$$\varphi_2 \equiv \Phi(\varphi_1) = 1 + \Delta(\xi_1) - \xi_1 \tag{2}$$

Fixed points of this map correspond to the periodic leap-frog activity:

$$\varphi = 1 + \Delta(\xi) - \xi \tag{3}$$

Since $\xi \equiv 1 + \varphi - \Delta(\varphi)$, this condition can be written in a more symmetric form

$$\varphi = \frac{\Delta(\varphi) + \Delta(\xi)}{2} \tag{4}$$

Taking into account the domain constraints $\xi \leq 1$, $\varphi \leq 1$, we also have

$$\Delta(\varphi) > \varphi, \; \Delta(\xi) < \xi \tag{5}$$

Conditions (4) and (5) are examined geometrically in Fig. 4B. Note that the synchronous firing solution $\{\varphi = 0^+, \xi = 1^-\}$ always satisfies the periodicity condition (4), if one assumes $\Delta(0^+) = \Delta(1^-) = 0$. If the inequality $\xi \leq 1$ is violated (i.e. when $\Delta(\varphi) < \varphi$), the cells fire sequentially, so their firing order does not alternate, while the violation of the condition $\Phi(\varphi) \leq 1$ (i.e. if $\Delta(\xi) > \xi$) indicates that the postsynaptic cell will emit more than two consecutive spikes.

Stability of the periodic leap-frog spiking depends on the value of the derivative of the phase map given by Eq. 1 at equilibrium:

$$\Phi'(\varphi) = [\Delta'(\xi) - 1][1 - \Delta'(\varphi)] \tag{6}$$

The fixed point will be stable if $|\Phi'(\varphi)| < 1$; an equivalent stability condition was derived in Ref. 1. The stability of synchronous firing is determined by an analogous map slope expression, with $\varphi = 0^+$ and $\xi = 1^-$ (Eq. 12 in Ref. 8). Since $\Delta'(1^-) \approx 0$ in the ML model (see Fig. 4B), the bifurcation from synchronous to leap-frog firing occurs when the slope $\Delta(0^+)$ becomes greater than 2, forcing φ to increase (and ξ to decrease) until the stability condition is satisfied. Thus, the characteristic sharp initial rise of $\Delta(\varphi)$ followed by a less steep increase at larger φ evident in Fig. 4B is essential for the transition from synchronous to leap-frog spiking. This feature corresponds to the characteristic dip to negative values in the phase transition return map noted in Ref. 1.

4. Effect of variation in coupling strength

Maran and Canavier[1] found that increasing the coupling strength causes the network to undergo a period-doubling cascade to chaos, whereby the synchronous dynamics seen for weak coupling transitions to leap-frog spiking in Fig. 1, followed by higher-period alternating-order states. Note that this period-doubling is readily explained by the map, Eq. (1), if one assumes a simple scaling between the coupling strength and the maximal amplitude of the STRC. In order to confirm this, we replaced the STRC with a simple quadratic function satisfying the leap-frog existence conditions established above, and "emulated" (artificially generated) the corresponding spike event sequence for different values of the amplitude of the quadratic STRC. The resulting bifurcation diagram was in good qualitative agreement with the bifurcation structure of the full ML network.

5. Concluding remarks: leap-frog spiking in 1-D models

Finally, we note that the above analysis can be used to show stable leap-frog spiking in even simpler 1-D models, for example in networks of quadratic integrate-and-fire neurons with a finite reset, and in networks of 1-D phase oscillators with continuous synaptic interaction.

Acknowledgments

This work was partially supported by the NSF grant DMS-0417416.

Appendix A. Morris-Lecar model parameters

We consider two identical Morris-Lecar neurons[11] with type-I excitability:[12]

$$C\frac{dV}{dt} = -\bar{g}_{Ca} m_\infty(V)(V - V_{Ca}) - \bar{g}_K w(V - V_K) - \bar{g}_L(V - V_L) - I_{app} - I_{syn}$$
$$\frac{dw}{dt} = \frac{w_\infty(V) - w}{\tau_\infty(V)} \tag{A.1}$$

$m_\infty(V) = [1 + \tanh((V + 12)/18)]/2$, $w_\infty(V) = [1 + \tanh((V + 8)/6)]/2$, $\tau_\infty(V) = 3/[2\cosh((V + 8)/12)]$, $C = 2\mu\text{F/cm}^2$, $I_{app} = -14\mu\text{A/cm}^2$, $V_{Ca} = 120\text{mV}$, $V_K = -84\text{mV}$, $V_L = -60\text{mV}$, $g_{Ca} = 4\text{mS/cm}^2$, $g_K = 8\text{mS/cm}^2$, $g_L = 2\text{mS/cm}^2$. Unperturbed limit cycle period is 45ms.

Cells are synaptically coupled by $I_{syn} = \bar{g}_{syn} s(t)(V - V_{inh})$ where $\bar{g}_{syn}$ is the peak synaptic conductance and $V_{inh} = -80\text{mV}$. Each synaptic gating variable $s_i(t)$ (i=1,2) is controlled by the presynaptic cell potential, $V_{j\neq i}$:

$$\frac{ds_i}{dt} = -\frac{s_i}{\tau_{syn}}\sigma(V_{th} - V_j) + \frac{1 - s_i}{\tau_\gamma}\sigma(V_j - V_{th}) \tag{A.2}$$

where $V_{th} = -3$ mV is the synaptic threshold, $\sigma(x) = [1 + \tanh(4x)]/2$, and $\tau_{syn} = 1$ ms and $\tau_\gamma = 0.2$ ms are the synaptic decay and rising time constants, respectively.

References

1. S. K. Maran SK and C. C. Canavier, *J. of Comput. Nerosci.* **24**, 037(2008).
2. J. A. White *et al.*, *J. of Comput. Neurosci.* **5**, 005 (1998).
3. X.-J. Wang and G. Buzsáki, *J. Neurosci.* **16**, 6402 (1996).
4. C. D. Acker, N. Kopell and J. A. White, *J. Comput. Neurosci.* **15**, 71 (2007).
5. A. T. Winfree, *The Geometry of Biological Time,* Springer (2001).
6. Y. Kuramoto, *Chemical Oscillations, Waves, and Turbulence,* Springer (1984).
7. E. M. Izhikevich and Y. Kuramoto, *Encycl. Math. Phys.* **5**, 448 (2006).

8. P. Goel and G. B. Ermentrout, *Physica D.* **163**, 191 (2002).
9. Ermentrout GB (1996). Neural Computation 8:979-1001.
10. E. Marder and R. L. Calabrese, *Physiol. Rev.* **76**, 687 (1996).
11. C. Morris and H. Lecar, *Biophys. J.* **35**, 193 (1981).
12. J. Rinzel and B. Ermentrout, *Methods in Neuronal Modeling, Eds. C. Koch and I. Segev*, MIT Press (1998).

A BOUNDARY INTEGRAL STRATEGY FOR THE LAPLACE-BELTRAMI-DIRICHLET PROBLEM ON THE SPHERE S^2

S. GEMMRICH* and N. NIGAM

Department of Mathematics and Statistics, McGill University, Montreal, QC H3A 2K6, Canada
**E-mail: gemmrich@math.mcgill.ca*

We present a boundary integral strategy to solve the Dirichlet problem for the Laplace-Beltrami operator on a subsurface of S^2. The approach is used to study vortex motion on the sphere in the presence of impenetrable boundaries. We validate the method by comparing our results to recent developments in the field.

Keywords: boundary integral equations; Laplace-Beltrami operator.

1. Introduction

Many boundary value problems can be reformulated and solved in terms of boundary integrals. Well known examples include problems from potential theory, acoustic a nd electromagnetic scattering theory as well as elasticity. The numerical treatment of such boundary integral approaches has led to fast and efficient algorithms. The purpose of this paper is to apply similar ideas to the context of the Laplace-Beltrami operator on the sphere S^2. The operator plays an important role in the study of vortex motion on the sphere.[1,2] Our boundary integral strategy extends some of the recent developments[3,4] in the study of such vortex motion in the presence of impenetrable boundaries on the sphere. The current methods rely on - via stereographic projection onto the complex plane - the explicit knowledge of certain conformal mappings associated with the geometric structure of the problem. Their use is therefore restricted to cases with simple geometries, whereas in principle the boundary integral strategy is applicable to a much wider class of settings. As a test case we focus on the Dirichlet problem for the Laplace-Beltrami operator on a submanifold of S^2. Previous preliminary results[5] obtained with our method were promising. For the present paper we chose numerical test cases comparable to the settings in Crowdy[3] and Kidambi and Newton.[4] We emphasize that the ideas are applicable to a wider class of problems than just the study of point vortex motion on the sphere. Further applications might include climate models which involve the Laplace-Beltrami equation. Bonner, Graham and Smyshlyaev[6] used a similar boundary integral approach for the Laplace-Beltrami-Helmholtz equation on a submanifold of the sphere to compute diffraction coefficients in high-frequency acoustic or electromagnetic scattering from conical objects.

The paper is organized as follows. In Section 2 we give a short description of the boundary value model used to study the motion of point vortices on the sphere. Section 3 outlines a boundary integral strategy to solve the correct Dirichlet problem. And finally, we present numerical results in Section 4 to validate our approach.

2. Point vortex motion on a sphere with impenetrable walls

We consider the motion of a single point vortex in a closed and simply connected submanifold Ω of the unit sphere S^2. We assume that a vortex of unit strength is located at a point $x_0 \in \Omega$ and denote the boundary of the submanifold by Γ. Due to the incompressibility of the fluid the flow is described by a stream function $\Psi(x_0, x)$ in the sense that $u = \nabla\Psi \times \mathbf{e}_r$, where u is the fluid velocity and $\mathbf{e}_r$ denotes the unit vector in the radial direction. The vorticity is then defined as $\boldsymbol{\omega} = \omega\,\mathbf{e}_r := \nabla \times u$ and we assume that the fluid motion is irrotational except at the vortex, which implies $\omega = 0$ except at x_0. Ψ is naturally assumed to be constant along Γ. This ensures that the boundary is in fact a streamline of the motion or in other words that the fluid can perfectly slip along Γ. Without loss of generality (in the simply connected case) we can set this constant to zero. We then see that Ψ is in fact the Green's function for the Laplace-Beltrami operator on the submanifold Ω, i.e.

$$-\Delta_S \Psi(x_0, x) = \delta(|x_0 - x|) \qquad \text{for all } x \in \Omega \tag{1}$$

$$\Psi(x_0, x) = 0 \qquad \text{for all } x \in \Gamma. \tag{2}$$

As in the planar case, we decompose Ψ into two terms

$$\Psi(x_0, x) = U(x_0, x) + v(x_0, x),$$

where $U(x_0, x) = -1/(4\pi)\log(1 - \langle x_0, x\rangle)$ is the Green's function for the Laplace-Beltrami operator on the sphere without solid boundaries. It satisfies

$$-\Delta_S U(x_0, x) = \delta(|x_0 - x|) - \frac{1}{4\pi}.$$

Note that the constant term on the right hand side is inevitable according to Gauss's theorem, which says that all feasible vorticity fields need to integrate to zero over the whole sphere, i.e. $\int_{S^2} \omega d\sigma = 0$. To guarantee that Ψ solves Eqs. (1) and (2) we then need to find v which satisfies

$$-\Delta_S\, v(x_0, x) = \frac{1}{4\pi} \qquad \text{for all } x \in \Omega \tag{3}$$

$$v(x_0, x) = -U(x_0, x) \qquad \text{for all } x \in \Gamma. \tag{4}$$

The streamlines of the fluid motion are then given by the level curves of Ψ. And the equations governing the motion of a particle are[4]

$$\dot{\theta} = \frac{1}{\sin\theta}\frac{\partial\Psi}{\partial\varphi} \quad \text{and} \quad \dot{\varphi} = -\frac{1}{\sin\theta}\frac{\partial\Psi}{\partial\theta},$$

where we introduced spherical coordinates $\theta \in [0, \pi]$ and $\varphi \in [0, 2\pi]$, denoting the polar angle measured from the positive z-axis and the counterclockwise azimuthal angle in the xy-plane measured from the positive x-axis respectively.

3. An indirect boundary integral approach

We now solve the boundary value problem specified in Eqs. (3) and (4) and choose to do so using an indirect single layer ansatz. This leads to an integral equation of the first kind. Other choices might lead to different integral equations of the first or second kind.[5] Let us now define the single layer operator

$$(\tilde{V}\sigma)(x) := \int_\Gamma U(x, y)\, \sigma(y)\, ds_y \quad \text{for } \ x \in \Omega. \tag{5}$$

Here, we assume that the density function σ belongs to an appropriate Sobolev space on the boundary Γ. For the moment it suffices to require $\sigma \in \mathrm{L}^2(\Gamma)$. We can then apply the Laplace-Beltrami operator to Eq. (5) and get for $x \notin \Gamma$:

$$-\Delta_S(\tilde{V}\sigma)(x) = \int_\Gamma -\Delta_S U(x, y)\, \sigma(y)\, ds_y = -\frac{1}{4\pi}\int_\Gamma \sigma(y)\, ds_y. \tag{6}$$

Next, it is essential to understand the behavior of $\tilde{V}\sigma(x)$ as x approaches the boundary Γ if we want to achieve the boundary constraint (4) . The following lemma[5] settles this question.

Lemma 3.1. *For $x_* \in \Gamma$ we have:*

$$(V\sigma)(x_*) := \lim_{\Omega \ni x \to x_*} (\tilde{V}\sigma)(x) = \int_\Gamma U(x_*, y)\,\sigma(y)\,ds_y$$

as a weakly singular integral and hence $(\tilde{V}\sigma)$ is continuous across Γ.

Hence, as a direct consequence of Eq. (6) and Lemma 3.1 we can solve the boundary value problem Eq. (3) and Eq. (4) as follows.

Lemma 3.2. *If the density σ solves the boundary integral equation*

$$(V\sigma)(x) = -U(x_0, x) =: g(x) \quad \textit{for } x \in \Gamma$$

subject to the constraint

$$\int_\Gamma \sigma\,ds = -1,$$

then $v := \tilde{V}\sigma$ solves Eq. (3) and Eq. (4).

In fact, in order to ensure solvability we look for v in the form

$$v = \tilde{V}\sigma + p,$$

for some Lagrange multiplier $p \in \mathbb{R}$ and require that the density solves the boundary integral equation

$$(V\sigma)(x) + p = g(x) \qquad \text{for } x \in \Gamma$$

subject to the constraint

$$\int_\Gamma \sigma\,ds = -1.$$

Rather than solving this directly we consider the following weak saddle point formulation. Find $\sigma \in L^2(\Gamma)$ and a real number p such that

$$\langle V\sigma, \xi\rangle + p\langle 1, \xi\rangle = \langle g, \xi\rangle \tag{7}$$

$$q\left(\langle 1, \sigma\rangle + 1\right) = 0 \tag{8}$$

for all $\xi \in L^2(\Gamma)$ and all real numbers q. Here, $\langle\cdot,\cdot\rangle$ denotes the L^2 inner product along Γ. We can relax the smoothness requirement on σ and replace $L^2(\Gamma)$ with an appropriate Sobolev space. This is subject to future work.

4. Numerical experiments

We now apply our strategy to four different subsurfaces of the sphere which were treated in either Crowdy[3] or Kidambi and Newton.[4] Our results coincide with the ones presented in those papers. In all cases we solve discrete versions of the equations (7) and (8). In fact, we partition the boundary into elements τ_i and approximate $L^2(\Gamma)$ by a piecewise constant test function. As a basis for the discrete space we chose the indicator functions of the elements $\varphi_i = \chi(\tau_i)$ and arrive at the linear system

$$\mathtt{V}\,\mathbf{s} + p\,\mathbf{h} = \mathbf{g} \tag{9}$$

$$\mathbf{s} \cdot \mathbf{h} = -1. \tag{10}$$

Here, $\mathbf{s}$ is the coefficient vector of σ with respect to the φ_i, $\mathbf{h}$ contains the information about the meshsize, i.e. $\mathbf{h}_i = \langle 1, \varphi_i \rangle$ and $\mathbf{g}$ encodes the information about the right hand side, $\mathbf{g}_i = \langle g, \varphi_i \rangle$. The entries of the system matrix are

$$\mathtt{V}_{ij} = \int_\Gamma \int_\Gamma U(x, x_0)\, \varphi_i(x_0) \varphi_j(x)\, ds_x ds_{x_0} = \int_{\tau_i} \int_{\tau_j} U(x, x_0)\, ds_x ds_{x_0}.$$

Note that the integrand function has a singularity, when x and x_0 coincide.

4.1. *Example 1 - spherical cap*

Let $\Omega = \{x(\varphi, \theta) \mid \theta < \theta_*\}$ be the spherical cap bounded by the polar angle θ_*, whose boundary is the circle $\Gamma = \{x(\varphi, \theta) \mid \theta = \theta_*\}$. We divide the circle into N elements τ_i of equal angular length $2\pi/N$ and use these as our mesh.

The fundamental solution for points along Γ (i.e. for $\theta = \theta_0 = \theta_*$) can be written as $U(\varphi, \theta, \varphi_0, \theta_0) = -1/(4\pi) \log\left\{(1 - \cos(\varphi - \varphi_0)) \sin^2(\theta_*)\right\}$. It is a straightforward calculation to show that the entries of $\mathtt{V}$ can be computed as follows.

$$\mathtt{V}_{ij} = -\frac{\sin^2(\theta_*)}{4\pi} I_{ji} - \frac{\log(\sin^2(\theta_*)) \sin^2(\theta_*)}{N^2},$$

where the double integrals I_{ji} are given by

$$I_{ji} = \int_{\tau_i} \int_{\tau_j} \log\left(1 - \cos(\varphi - \varphi_0)\right)\, d\varphi d\varphi_0.$$

We note that the matrix $\mathtt{V}$ is a symmetric Toeplitz matrix and hence it is sufficient to compute its first row. The integrand function is singular when the arguments coincide. In those cases we separate the singularity according to $\log\{1 - \cos(t)\} = \log(t^2) + \log(1/2 - t^2/4! + t^4/6! + \ldots)$. The first term can be integrated exactly and since the second term is no longer singular any standard quadrature method will work to approximate its integral. The other data vectors in the saddle point problem (9) and (10) can be computed in a straight forward fashion. Figure 1 shows the computed solution for the spherical cap bounded by the polar angle $\theta_* = \pi/3$ with a vortex located at $\theta_0 = \pi/6$, $\varphi_0 = \pi/2$.

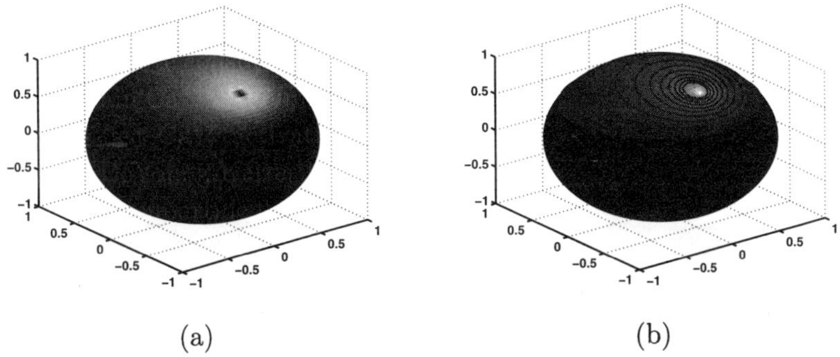

Fig. 1. polar cap: $\theta_* = \pi/3$, vortex position: $\theta_0 = \pi/6$, $\varphi_0 = \pi/2$, (a) solution to boundary value problem, (b) contour lines.

4.2. *Example 2 - longitudinal wedge*

We consider the wedge $\Omega = \{x(\varphi, \theta) \mid 0 < \varphi < \alpha\}$, for some angle $\alpha \in (0, 2\pi)$. Our mesh consists of a uniform partition into $2N$ elements of the boundary Γ. The matrix $\mathtt{V}$ is symmetric and has the following block structure

$$\mathtt{V} = \left(\begin{array}{c|c} \mathtt{B} & \mathtt{C} \\ \hline \mathtt{C}^t & \mathtt{B} \end{array}\right),$$

where $\mathtt{B}$ is the symmetric Toeplitz matrix which corresponds to the case when both elements of integration share the same azimuthal angle and can be computed using the ideas from the previous section. $\mathtt{C}$ includes the matrix entries for which the elements of integration differ in the azimuthal angle. In this case the only singular integral to compute is when both integration elements touch either the north or south pole. Those two cases are symmetric and we treat the first one here. Simple calculations show that the corresponding matrix entry can be computed by

$$V_{1,2N} = -\frac{1}{4\pi} \int_0^{\frac{\pi}{N}} \int_0^{\frac{\pi}{N}} \log\left\{1 - \cos(\alpha)\sin(t)\sin(s) - \cos(t)\cos(s)\right\} \, dsdt,$$

where the integrand is only singular for $t = 0$ and $s = 0$. The argument of the logarithm has the following asymptotic behavior.

$$1 - \cos(\alpha)\sin(t)\sin(s) - \cos(t)\cos(s) = -\cos(\alpha)ts + \frac{t^2}{2} + \frac{s^2}{2} + \text{h.o.t.}$$

Again we can separate the singularity of the integrand into a term that can be integrated exactly and a nonsingular term. Figure 2 shows the results for a wedge of width $\alpha = \pi/3$ and a vortex located at $\theta_0 = \pi/4$, $\varphi_0 = \pi/4$.

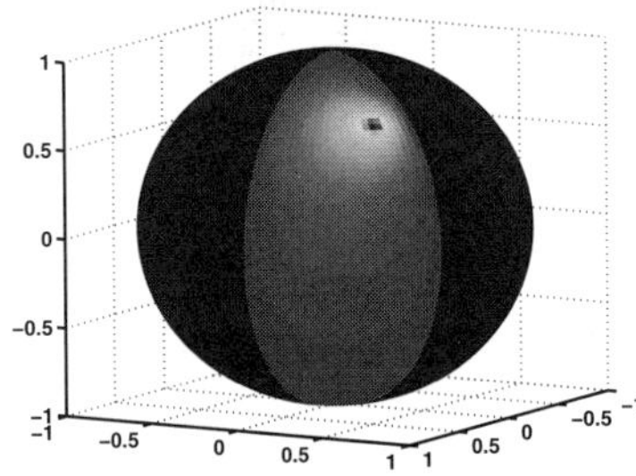

(a)

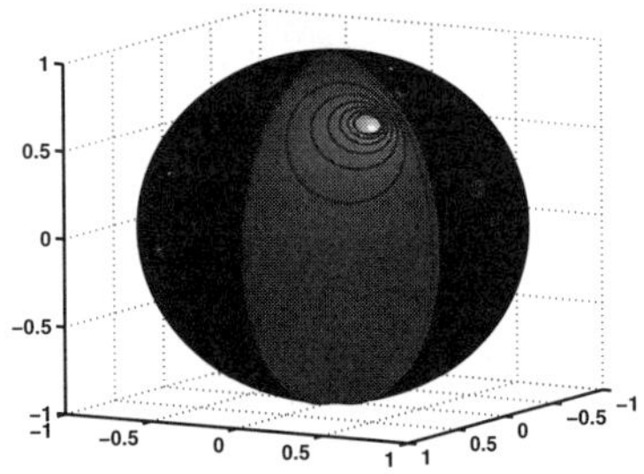

(b)

Fig. 2. longitudinal wedge: $\alpha = \pi/3$, vortex position: $\theta_0 = \pi/4$, $\varphi_0 = \pi/4$, (a) solution to boundary value problem, (b) contour lines.

4.3. *Example 3 - half-longitudinal wedge*

We consider the half wedge $\Omega = \left\{x(\varphi,\theta) \,|\, 0 < \varphi < \alpha,\ \frac{\pi}{2} < \theta < \pi\right\}$ for some angle $\alpha \in (0, 2\pi)$. Our mesh consists of a semi–uniform discretization of the boundary Γ, where the mesh sizes were chosen constant on each of the three parts of the boundary. The matrix $\mathtt{V}$ is again symmetric and has the following block structure

$$\mathtt{V} = \left(\begin{array}{c|c|c} \mathtt{B}_1 & \mathtt{C}_{12} & \mathtt{C}_{13} \\ \hline {\mathtt{C}_{12}}^t & \mathtt{B}_2 & \mathtt{C}_{23} \\ \hline {\mathtt{C}_{13}}^t & {\mathtt{C}_{23}}^t & \mathtt{B}_3 \end{array}\right).$$

Here, the $\mathtt{B}_i$ correspond to the entries where both elements of integration lie on the same piece of the boundary and the $\mathtt{C}_{ij}$ correspond to entries of cross integration. In case of a singular integrand, we separate the singularity according to the ideas presented in the previous two sections. Figure 3 shows the results for a half-wedge of width $\alpha = \pi/2$ and a vortex located at $\theta_0 = 3\pi/4$, $\varphi_0 = \pi/6$.

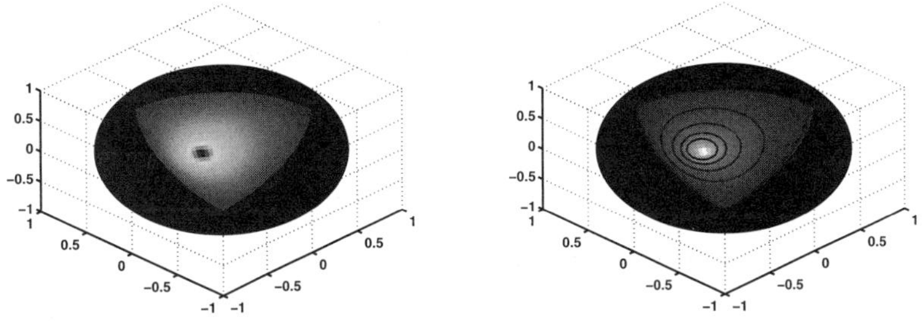

(a) (b)

Fig. 3. half longitudinal wedge: $\alpha = \pi/2$, vortex position: $\theta_0 = 3\pi/4$, $\varphi_0 = \pi/6$, (a) solution to boundary value problem, (b) contour lines.

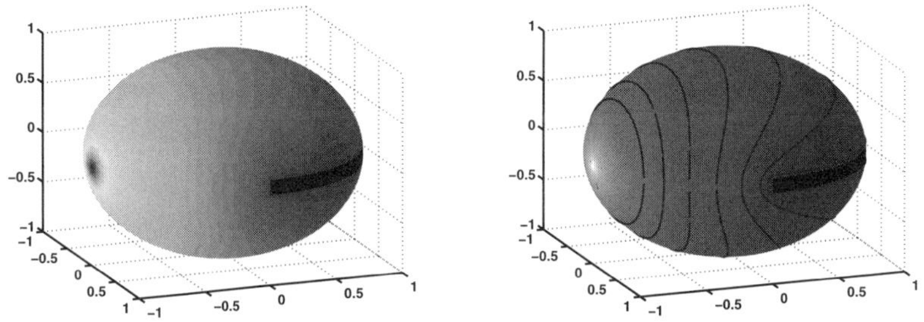

(a) (b)

Fig. 4. rectangular wall with gap of length π and thickness $\delta = \pi/50$, vortex position: $\theta_0 = \pi/2$ and $\varphi_0 = \pi$, (a) solution to boundary value problem, (b) contour lines.

4.4. *Example 4 - gap in a rectangular wall*

For some small $\delta > 0$ we consider the rectangular 'wall' $\Omega = \{x(\varphi, \theta) \,|\, \varphi \in [0, \pi],\, \theta \in [\frac{\pi}{2} - \delta, \frac{\pi}{2} + \delta]\}$, which is aligned with the equator and leaves a gap of length π. The mesh is given by a semi–uniform discretization of the rectangular boundary Γ. We set the mesh sizes constant on each of the four boundary parts. As before, possible singularities in the integrand functions are separated and integrated exactly. Figure 4 shows the results for a rectangular wall of thickness $\delta = \pi/50$ and a vortex located at $\theta_0 = \pi/2$ and $\varphi_0 = \pi$.

References

1. Y. Kimura, *R. Soc. Lond. Proc. Ser. A Math. Phys. Eng. Sci.* **455**, 245 (1999).
2. D. Crowdy and M. Cloke, *Phys. Fluids* **15**, 22 (2003).
3. D. Crowdy, *Phys. Fluids* **18** (2006).
4. R. Kidambi and P. K. Newton, *Phys. Fluids* **12**, 581 (2000).
5. S. Gemmrich, N. Nigam and O. Steinbach, *Proceedings of the Third Abel Symposium* (to appear 2008).
6. B. D. Bonner, I. G. Graham and V. P. Smyshlyaev, *SIAM J. Numer. Anal.* **43**, 1202 (2005).

NON-SPATIAL WHOLE CELL MODELS OF GLOBAL CALCIUM RESPONSES THAT ACCOUNT FOR HETEROGENEOUS DOMAIN CALCIUM CONCENTRATIONS

GEORGE S. B. WILLIAMS, MARCO A. HUERTAS, and GREGORY D. SMITH*

Dept. of Applied Science, The College of William and Mary, Williamsburg, VA 23187, USA
**E-mail: greg@as.wm.edu*
http://www.as.wm.edu/Faculty/Smith.html

M. SALEET JAFRI

Dept. of Bioinformatics and Computational Biology, George Mason University, Manassas, Virginia 20110, USA

ERIC A. SOBIE

Dept. of Pharmacology and Systems Therapeutics, Mount Sinai School of Medicine, New York, New York 10029, USA

A limitation of most whole cell models to date is the assumption that intracellular Ca^{2+} channels are globally coupled by a continuously stirred bulk cytosolic $[Ca^{2+}]$, when in fact open intracellular Ca^{2+} channels experience elevated domain $[Ca^{2+}]$. Such heterogeneous local $[Ca^{2+}]$ can be modeled using a $2N+2$-compartment model that includes bulk cytosolic and luminal $[Ca^{2+}]$ and $2N$ compartments representing cytosolic and luminal Ca^{2+} domains associated with N stochastically gating intracellular channels. We have introduced an alternative whole cell model formulation that solves a system of advection-reaction equations for the probability density of cytosolic and luminal domain $[Ca^{2+}]$ jointly distributed with channel state. This probability density formulation and an associated moment closure approach have been used to create computationally efficient models of local control of Ca^{2+}-induced Ca^{2+} release in ventricular cardiac myocytes.

Keywords: whole cell model; local calcium signaling; calcium domain; stochastic gating; markov chain; probability density; moment closure.

1. Introduction

Whole cell models of intracellular Ca^{2+} signaling are usually Hodgkin-Huxley-like systems of nonlinear ordinary differential equations (ODEs). Such models have played a role in understanding endoplasmic reticulum (ER) Ca^{2+} excitability and oscillations in gonadotrophs, rat basophilic leukemia cells, cardiac myocytes, and many other cell types (see Ref. 1 for review). For example, two-compartment whole cell models often take the form

$$\frac{dc_{cyt}}{dt} = J_{rel} + J_{leak} - J_{pump} \tag{1}$$

$$\frac{dw}{dt} = \frac{w_\infty - w}{\tau_w} \tag{2}$$

$$\frac{dc_{er}}{dt} = -\frac{1}{\lambda_{er}} \left[J_{rel} + J_{leak} - J_{pump}\right] \tag{3}$$

where λ_{er} is the ER-to-cytosolic effective volume ratio that accounts for the binding capacity of Ca^{2+} buffers. The three fluxes influencing the $[Ca^{2+}]$ in the cytosol (c_{cyt}) and ER (c_{er}) include: Ca^{2+} release via inositol 1,4,5-trisphosphate (IP_3) receptors (IP_3Rs),

$$J_{rel} = v_{rel} f_{open} \left(c_{er} - c_{cyt}\right), \tag{4}$$

a passive leak from the ER to cytosol,

$$J_{leak} = v_{leak} \left(c_{er} - c_{cyt}\right), \tag{5}$$

and Ca^{2+} reuptake via SERCA-type Ca^{2+}-ATPases,

$$J_{pump} = \frac{v_{pump} c_{cyt}^2}{c_{cyt}^2 + k_{pump}^2}. \tag{6}$$

In this traditional whole cell model, w is a Hodgkin-Huxley-like gating variable satisfying a first order kinetic equation (Eq. 2). For example, in the Li-Rinzel reduction of the DeYoung-Keizer IP_3R model,[2,3] the variable w represents the fraction of IP_3Rs that are *not* inactivated, $w_\infty(c_{cyt})$ and $\tau_w(c_{cyt})$ are both functions of the cytosolic $[Ca^{2+}]$, and $f_{open}(w, c_{cyt})$ is the fraction of open IP_3Rs. By numerically integrating conventional whole cell models such as Eqs. 1–6, one may simulate Ca^{2+} release and reuptake by IP_3-sensitive intracellular Ca^{2+} stores.

While considerable insight has been obtained through the analogy of plasma membrane electrical excitability and ER Ca^{2+} excitability, the disparity between electrical length scales (100–200 μm) and the range of action of intracellular Ca^{2+} (i.e., a *chemical* length scale of 1–2 μm) suggests that some aspects of the analogy are strained.[4–7] In particular, the ODE for the gating variable representing IP_3R inactivation (Eq. 2) is an average rate equation that is derived by assuming a large number of intracellular Ca^{2+} channels are globally coupled via bulk cytosolic Ca^{2+} (c_{cyt}).[8] While it is true that plasma membrane ion channels in an electrotonically compact cell experience essentially the same time-course of membrane voltage, intracellular Ca^{2+} channels experience radically different local [Ca^{2+}], even during global Ca^{2+} responses, and clusters of IP_3Rs and/or ryanodine receptors (RyRs) are in fact only *locally* coupled via the buffered diffusion of intracellular Ca^{2+}. For this reason, the Hodgkin-Huxley-style average rate equations such as Eq. 2 are not always appropriate.

2. Many-compartment Monte Carlo model

A whole cell model of local and global Ca^{2+} dynamics with luminal and cytosolic Ca^{2+} domains associated with each IP_3R can be simulated using a Monte Carlo approach where the domains and bulk Ca^{2+} concentrations

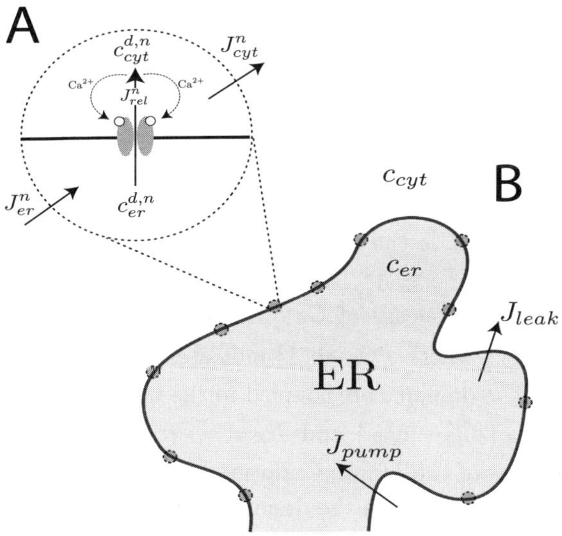

Fig. 1. Diagram of a $2N+2$-compartment Monte Carlo model. A) Local Ca^{2+} signaling near the nth intracellular Ca^{2+} channel includes diffusion (J^n_{er} and J^n_{cyt}) and release (J^n_{rel}) fluxes that are functions of cytosolic and luminal domain Ca^{2+} ($c^{d,n}_{cyt}$ and $c^{d,n}_{er}$). B) The restorative flux from SERCA pumps (J_{pump}) and leak from the ER to the cytosol (J_{leak}) are functions of the bulk cytosolic and bulk ER [Ca^{2+}] (c_{cyt} and c_{er}). Reproduced with permission from Ref. 9.

solve a large number of ODEs coupled to Markov chains representing the stochastically gating channels. Figure 1 shows a diagram of the components and fluxes of a representative model with $2N+2$ compartments. Each IP_3R is described by an M-state Markov chain with an infinitesimal generator matrix (sometimes called the Q-matrix) that collects the rate constants of the single channel model,

$$Q\left(c_{cyt}^{d}, c_{er}^{d}\right) = K^{-} + \left(c_{cyt}^{d}\right)^{\eta} K_{cyt}^{+} + \left(c_{er}^{d}\right)^{\eta} K_{er}^{+} \tag{7}$$

where the off-diagonal elements of K^{-} are Ca^{2+}-independent transition rates and K_{cyt}^{+} and K_{er}^{+} include the association rate constants for the transitions mediated by cytosolic and luminal domain Ca^{2+}. A four-state Markov chain model can represent the dynamics of fast Ca^{2+} activation and slow Ca^{2+} inactivation similar to the well-known DeYoung-Keizer IP_3R model.[2]

The time evolution of $[Ca^{2+}]$ in each bulk compartment and domain are modeled by a set of $2N+2$ ordinary differential equations,

$$\frac{dc_{cyt}}{dt} = J_{cyt}^{T} + J_{leak} - J_{pump} \tag{8}$$

$$\frac{dc_{er}}{dt} = \frac{1}{\lambda_{er}}\left(-J_{er}^{T} - J_{leak} + J_{pump}\right) \tag{9}$$

$$\frac{dc_{cyt}^{d,n}}{dt} = \frac{1}{\lambda_{cyt}^{d}}\left(J_{rel}^{n} - J_{cyt}^{n}\right) \tag{10}$$

$$\frac{dc_{er}^{d,n}}{dt} = \frac{1}{\lambda_{er}^{d}}\left(-J_{rel}^{n} + J_{er}^{n}\right), \tag{11}$$

where $1 \leq n \leq N$, N is the number of IP_3Rs, $J_{cyt}^{T} = \sum_{n=1}^{N} J_{cyt}^{n}$, $J_{er}^{T} = \sum_{n=1}^{N} J_{er}^{n}$, and λ_{er}, $\lambda_{cyt}^{d} = \Lambda_{cyt}^{d}/N$, and $\lambda_{er}^{d} = \Lambda_{er}^{d}/N$ are effective volume ratios. Note that the release of Ca^{2+} through IP_3Rs is given by $J_{rel}^{n} = \gamma^{n} v_{rel}\left(c_{er}^{d,n} - c_{cyt}^{d,n}\right)$ where $\gamma^{n} = \{0, 1\}$ indicates whether the nth channel is open or closed. The domains are coupled to the bulk compartments via the fluxes $J_{cyt}^{n} = v_{cyt}\left(c_{cyt}^{d,n} - c_{cyt}\right)$ and $J_{er}^{n} = v_{er}\left(c_{er} - c_{er}^{d,n}\right)$. Figure 2 shows that in simulations of the Monte Carlo model (Eqs. 8–11) the choice of N has profound consequences on the dynamics of the bulk cytosolic $[Ca^{2+}]$.

3. Probability density formulation

The probability density approach is an alternative to Monte Carlo simulation that begins by defining continuous multivariate probability density functions for the cytosolic domain $[Ca^{2+}]$ ($\tilde{c}_{cyt}^{d}$) and luminal domain $[Ca^{2+}]$

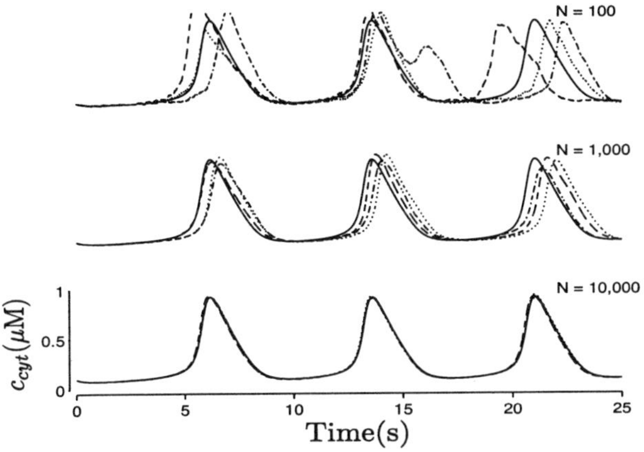

Fig. 2. Simulated oscillations of bulk cytosolic $[Ca^{2+}]$ (c_{cyt}) with increasing numbers of IP_3Rs (N=100; 1,000; 10,000). The *dotted, dashed,* and *dot-dashed lines* correspond to three distinct Monte Carlo simulations. The *solid line* is the probability density result (replotted in each panel). Reproduced with permission from Ref. 9.

($\tilde{c}^d_{er}$) jointly distributed with IP_3R state, $\tilde{S}(t)$,[10]

$$\rho^i(c^d_{cyt}, c^d_{er}, t)\, dc^d_{cyt}\, dc^d_{er} = \mathsf{P}\{c^d_{cyt} < \tilde{c}^d_{cyt} < c^d_{cyt} + dc^d_{cyt} \text{ and } c^d_{er} < \tilde{c}^d_{er} < c^d_{er} + dc^d_{er} \text{ and } \tilde{S}(t) = i\}. \quad (12)$$

For these multivariate probability densities to be consistent with the dynamics of the Monte Carlo model in the previous section, they must solve the following system of advection-reaction equations,

$$\frac{\partial \rho^i}{\partial t} = -\frac{\partial}{\partial c_{cyt}}[f^i_{cyt}\,\rho^i] - \frac{\partial}{\partial c_{er}}[f^i_{er}\,\rho^i] + [\boldsymbol{\rho} Q]^i, \quad (13)$$

where $1 \le i \le M$, Q is the generator matrix for the M-state IP_3R model (Eq. 7), $\boldsymbol{\rho}(c_{jsr}, t) = (\rho^1, \rho^2, \cdots, \rho^M)$, and $[\boldsymbol{\rho} Q]^i$ is the ith element of the vector-matrix product $\boldsymbol{\rho} Q$. The reaction term $[\boldsymbol{\rho} Q]^i$ describe changes in probability due to IP_3R stochastic gating, while the advection rates $f^i_{cyt}(c^d_{cyt}, c^d_{er})$ and $f^i_{er}(c^d_{cyt}, c^d_{er})$ describe the deterministic aspect of the time-evolution of the domain Ca^{2+} when the IP_3R is in state i, for example,

$$f^i_{cyt}(c^d_{cyt}, c^d_{er}) = \frac{1}{\Lambda^d_{cyt}}\left[\gamma^i v^T_{rel}\left(c^d_{er} - c^d_{cyt}\right) - v^T_{cyt}\left(c^d_{cyt} - c_{cyt}\right)\right]. \quad (14)$$

In the probability density formulation, c_{cyt} and c_{er} satisfy ODEs similar in form to Eqs. 8 and 9 with J^T_{cyt} and J^T_{er} given by functionals of the $\rho^i(c^d_{cyt}, c^d_{er}, t)$ solving Eq. 13.

4. Probability density simulations of local and global Ca^{2+} responses

Figure 2 shows Monte Carlo simulations converging to the probability density result as the number of IP_3Rs increases. When the Monte Carlo calculation includes relatively few IP_3Rs ($N = 100$), the stochastic gating of the channels leads to fluctuations in the oscillatory dynamics of bulk cytosolic $[Ca^{2+}]$ that decrease when N is increased to 1,000. In the Monte Carlo simulation with $N = 10{,}000$ channels, the dynamics of the bulk cytosolic $[Ca^{2+}]$ is very similar to the probability density result (compare *broken* and *solid lines*). In addition to demonstrating the validity of the probability density approach to whole cell modeling of local and global Ca^{2+} dynamics, simulations similar to Fig. 2 have shown that the probability density approach becomes computationally more efficient than the Monte Carlo simulation long before a physiologically realistic number of IP_3Rs are included.[9]

The probability density approach is, of course, not limited to whole cell Ca^{2+} responses mediated by IP_3Rs, but can be generalized to Markov chain models of intracellular Ca^{2+} channels of arbitrary complexity that include cytosolic and luminal Ca^{2+} regulation, and even clusters of spatially co-localized Ca^{2+}-regulated Ca^{2+} channels. Recently the probability density approach has been benchmarked in the context of mechanistic local control models of Ca^{2+}-induced Ca^{2+} release (CICR) that reproduce high-gain graded release by including the stochastic gating of a large number of Ca^{2+} release units (CaRUs).[11] In such models, each CaRU is composed of one or more L-type Ca^{2+} channels interacting with a cluster of RyRs through changes in $[Ca^{2+}]$ in a small diadic subspace between the sarcolemmal and sacroplasmic reticulum (SR) membranes, and local changes in junctional SR $[Ca^{2+}]$ are often important for realistic termination and refractoriness of localized Ca^{2+} release.[12–16] We have found that local control of CICR can be efficiently simulated using an advection-reaction system similar in form to Eq. 13, especially when rapid equilibrium of diadic subspace $[Ca^{2+}]$ with juctional SR and bulk myoplasmic $[Ca^{2+}]$ allows the use of univariate probability densities for the joint distribution of junctional SR Ca^{2+} and CaRU state.[11]

Beginning with this univariate probability density approach, further computational efficiency has been attained using a moment closure technique.[10] This method begins with a derivation of a system of ODEs describing the time-evolution of the moments of the univariate probability density functions for junctional SR $[Ca^{2+}]$ jointly distributed with CaRU state.

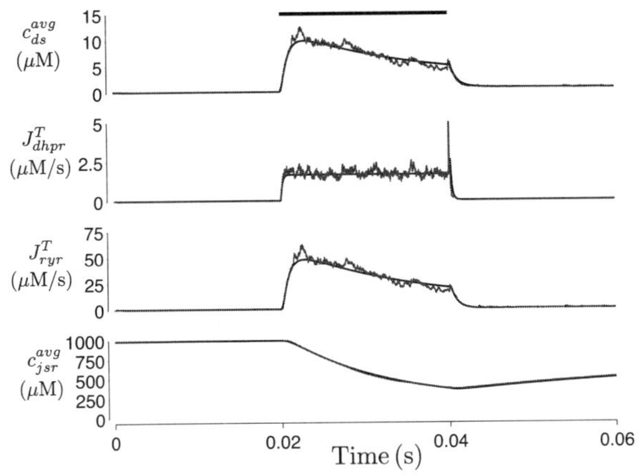

Fig. 3. The response of a whole cell ventricular cardiac myocyte model during a 20 ms step depolarization from a holding potential of −80 mV to −10 mV (*bar*) with the Monte Carlo and moment closure results indicated as a *grey line* and *black line*, respectively. From top to bottom: average diadic subspace [Ca^{2+}], total Ca^{2+} flux via the L-type Ca^{2+} channels (J_{dhpr}^{T}), total Ca^{2+}-induced Ca^{2+} release flux (J_{ryr}^{T}), and average junctional SR [Ca^{2+}]. The Monte Carlo simulation used $N = 1000$ CaRUs. Reproduced with permission from Ref. 10.

This open system of ODEs is then closed using an algebraic relationship that expresses the third moment of junctional SR [Ca^{2+}] in terms of the first and second moments. In this manner, the partial differential equations for the univariate probability densities are replaced with ODEs describing the time-evolution of the moments of these distributions. Benchmark simulations indicate that the moment closure technique for local control models of CICR in cardiac myocytes is nearly 10,000 times more computationally efficient than corresponding Monte Carlo simulations, while leading to nearly identical results (see Fig. 3).

5. Conclusions and future work

Although the computational efficiency of the moment closure technique in the context of local control models of CICR is exciting, the relative merits of Monte Carlo, probability density, and moment closure methods are in general model dependent. For example, the run time required for the Monte Carlo simulations such as Fig. 3 is a linear function of the number of CaRUs (N). Similarly, the computational efficiency of the univariate probability density calculation of Fig. 3 scales linearly with the number of Ca^{2+} release unit states (M) and the number of mesh points used to discretize the junctional SR [Ca^{2+}].[11] Because the moment closure approach results in $2+3M$ ODEs (bulk myoplasmic [Ca^{2+}], network SR [Ca^{2+}], and three moments for each CaRU state), the computational requirements of the moment closure approach is expected to scale linearly with M. That is, when Ca^{2+} release units are defined compositionally from single channel models of L-type Ca^{2+} channels and RyRs, the resulting large number of CaRU states will reduce the computational advantage of the moment closure approach relative to Monte Carlo. For this reason, the development of largeness tolerance and largeness avoidance techniques[17] for compositionally defined CaRU models is an important goal for further research in non-spatial whole cell models of global Ca^{2+} responses that account for heterogeneous domain calcium concentrations.

Acknowledgments

This conference proceedings briefly summarizes results of three research articles.[9–11] This material is based upon work supported by the National Science Foundation under Grants No. 0133132 and 0443843.

References

1. G. D. Smith, J. Pearson and J. Keizer, Modeling intracellular Ca^{2+} waves and sparks, in *Computational Cell Biology*, eds. C. P. Fall, E. S. Marland, J. M. Wagner and J. J. Tyson (Springer-Verlag, 2002) pp. 200–232.
2. G. W. DeYoung and J. Keizer, *Proc Natl Acad Sci USA* **89**, 9895(Oct 1992).
3. Y. X. Li and J. Rinzel, *J Theor Biol* **166**, 461(Feb 1994).
4. E. Neher, *Exp Brain Res* **14**, 80 (1986).
5. N. L. Allbritton, T. Meyer and L. Stryer, *Science* **258**, 1812(Dec 1992).
6. M. Naraghi and E. Neher, *J Neurosci* **17**, 6961(Sep 1997).
7. G. D. Smith, L. Dai, R. Muira and A. Sherman, *SIAM J Appl Math* **61**, 1816 (2001).
8. G. D. Smith, Modeling the stochastic gating of ion channels, in *Computational Cell Biology*, eds. C. P. Fall, E. S. Marland, J. M. Wagner and J. J. Tyson (Springer-Verlag, 2002) pp. 291–325.
9. G. S. B. Williams, E. J. Molinelli and G. D. Smith, *J Theor Biol* **253**, 170 (2008).
10. G. S. B. Williams, M. A. Huertas, E. A. Sobie, M. S. Jafri and G. D. Smith, *Biophys J* (2008), in press.
11. G. S. B. Williams, M. A. Huertas, E. A. Sobie, M. S. Jafri and G. D. Smith, *Biophys J* **92**, 2311 (2007).
12. M. D. Stern, L. S. Song, H. Cheng, J. S. Sham, H. T. Yang, K. R. Boheler and E. Ríos, *J Gen Physiol* **113**, 469(Mar 1999).
13. J. J. Rice, M. S. Jafri and R. L. Winslow, *Biophys J* **77**, 1871 (1999).

14. M. D. Stern, *Biophys J* **63**, 497 (1992).
15. E. A. Sobie, K. W. Dilly, J. dos Santos Cruz, W. J. Lederer and M. S. Jafri, *Biophys J* **83**, 59 (2002).
16. J. L. Greenstein and R. L. Winslow, *Biophys J* **83**, 2918 (2002).
17. H. DeRemigio, P. Kemper, M. D. LaMar and G. D. Smith, *Physical Biology* (2008), in press.

STABLE AND ACCURATE OUTGOING WAVE FILTERS FOR ANISOTROPIC AND NONLOCAL WAVES

AVY SOFFER

Department of Mathematics, Rutgers University
Piscataway, NJ 08854, USA

CHRIS STUCCHIO

Courant Institute of Mathematical Sciences, New York University
NY, NY, 10012-1185, USA
http://cims.nyu.edu/~stucchio/

The Perfectly Matched Layer (PML) is currently the mainstay of absorbing boundary conditions. For some anisotropic wave equations the PML is exponentially unstable in time. We present in this work a new method of open boundaries, the phase space filter, which is stable for all wave equations.

Outgoing waves can be are waves located near the boundary of the computational domain with group velocities pointing out. Phase space filtering involves periodically removing only outgoing waves from the solution, leaving non-outgoing waves unchanged. We apply this method to the Euler equations (linearized about a jet flow), Maxwell equations in a birefringent medium and the quasi-geostrophic equations.

1. Introduction

We consider numerical approximations of linear wave equations:

$$\boldsymbol{u}_t(\boldsymbol{x},t) = \mathcal{H}\boldsymbol{u}(\boldsymbol{x},t) \tag{1}$$

$\boldsymbol{u} : \mathbb{R}^N \to \mathbb{R}^n$ (or $\mathbb{R}^N \to \mathbb{C}^n$) is a vector-valued wave-field and $\mathcal{H} = \mathcal{H}(i\nabla)$ is a skew-adjoint linear differential operator. Eq. (1) has plane wave solutions $e^{i(\boldsymbol{k}\cdot\boldsymbol{x}-\omega_j(\boldsymbol{k})t)}\boldsymbol{d}_j(\boldsymbol{k})$ with $\boldsymbol{d}_j(\boldsymbol{k})$ the j'th eigenvector of $\mathcal{H}(\boldsymbol{k})$ and $\omega_j(\boldsymbol{k})$ the j-th eigenvalue of $\mathcal{H}(\boldsymbol{k})$ in the frequency domain. If we are only interested in $\boldsymbol{u}(x,t)$ in a finite region $B \subset \mathbb{R}^N$, it simplifies the computations to solve Eq. (1) only on this region. Steps must be taken to prevent spurious reflection from the artificial boundary, which is the topic of this article.

Exact transparent boundary conditions can be constructed, but can be difficult to work with.[1–4] The PML is a more versatile,[5] but for some anisotropic wave equations it can be exponentially unstable in time[5–7] which makes it inaccurate*.

The Time Dependent Phase Space Filter (TDPSF) is a new approach to the problem of open boundaries,[10–13] and in this work we apply it the problem of anisotropic waves. The key idea in this approach is that outgoing waves live in certain regions of *phase space*; by filtering these regions, outgoing waves can be removed before they reach the computational boundary.

The phase space projections originally used were based on the Gaussian windowed Fourier transform.[10–12] Here, we extend the work of[10–13] to more general wave equations, and simultaneously simplify the method. We begin by briefly reviewing the dynamics of linear waves.

1.1. *Outgoing waves*

To pin down notation, we review wave propagation. Recall that $\mathcal{H}(i\nabla)$ is skew-adjoint, i.e. for each $\boldsymbol{k}$, $\mathcal{H}(\boldsymbol{k})$ is a skew-adjoint matrix (with complex entries). $\mathcal{H}(\boldsymbol{k})$ can be diagonalized by the matrix $D = D(\boldsymbol{k})$, which is the (unitary) matrix having j'th row equal to $\boldsymbol{d}_j(\boldsymbol{k})$:

$$\mathcal{H} = D^{\dagger} \begin{bmatrix} i\omega_1(\boldsymbol{k}) \dots & 0 & \\ \dots & i\omega_j(\boldsymbol{k}) \dots & \\ 0 & \dots & i\omega_n(\boldsymbol{k}) \end{bmatrix} D \tag{2}$$

The function $\boldsymbol{d}_j(\boldsymbol{k})e^{i(\boldsymbol{k}\cdot x - \omega_j(\boldsymbol{k})t)}$ is a plane wave solution to Eq. (1). The operator $e^{\mathcal{H}t}$ is the propagator for (1), i.e. $e^{\mathcal{H}t}\boldsymbol{u}(x,0) = \boldsymbol{u}(x,t)$. In practice, we compute this with Fast Fourier Transforms, i.e. $\mathrm{FFT}^{-1}\exp(\mathcal{H}(\boldsymbol{k})t)\mathrm{FFT}\boldsymbol{u}(x,0)$.

Consider an initial condition $\boldsymbol{u}(x,0) = e^{i\boldsymbol{k}_0 x}\boldsymbol{d}_j(\boldsymbol{k}_0)g(\boldsymbol{x}-\boldsymbol{a})$, with $g(\boldsymbol{x})$ smooth and well localized (e.g. a gaussian). Stationary phase shows that:

$$\begin{aligned} \boldsymbol{u}(x,t) &= e^{\mathcal{H}(\boldsymbol{k})t}\boldsymbol{d}_j(\boldsymbol{k}_0)\hat{g}(\boldsymbol{k}-\boldsymbol{k}_0) \\ &= \exp([\omega_j(\boldsymbol{k}_0) + \nabla_k\omega_j(\boldsymbol{k}_0) + O((\boldsymbol{k}-\boldsymbol{k}_0)^2)]t)\boldsymbol{d}_j(\boldsymbol{k}_0)e^{i\boldsymbol{k}\cdot\boldsymbol{a}}\hat{g}(\boldsymbol{k}-\boldsymbol{k}_0) \\ &\approx e^{i(\boldsymbol{k}_0 x - \omega_j(\boldsymbol{k}_0)t)}g_t(\boldsymbol{x}-\boldsymbol{a}-\nabla_k\omega_j(\boldsymbol{k}_0)t) \end{aligned} \tag{3}$$

The envelope $g_t(x)$ disperses due to the $O((\boldsymbol{k}-\boldsymbol{k}_0)^2)$ term. Thus, wavepackets of the form $\boldsymbol{u}(x,0) = e^{i\boldsymbol{k}_0 x}\boldsymbol{d}_j(\boldsymbol{k}_0)g(\boldsymbol{x}-\boldsymbol{a})$ propagate along the trajectory $\boldsymbol{a}+\nabla_k\omega_j(\boldsymbol{k}_0)t$ while spreading out. The TDPSF algorithm consists of identifying and removing wave components with outgoing trajectories (and only outgoing trajectories) before they reach the computational boundary.

*It can also be polynomially unstable in time[8] due to problems near $\boldsymbol{k} = 0$ which is unrelated to anisotropy. This issue has been resolved.[9]

2. The TDPSF Algorithm

The TDPSF algorithm involves solving Eq. (1) on the extended region $[-L - w, L + w]^N$ with any accurate interior solver, with $[-L, L]^N$ the region of interest, and the extra space a filtering buffer.

We assume that $\boldsymbol{u}(x, 0)$ has small high frequency components, as well as a more technical assumption that the frequency content of $\widehat{\boldsymbol{u}}_0(\boldsymbol{k})$ is localized away from the regions where the group velocity turns around (see Section 2.1). This requirement is necessary due to the Heisenberg uncertainty principle. In what follows, $v_{\max}$ is the largest relevant group velocity:

Algorithm 1. ***TDPSF Propagation***

Input:

- *The dispersion relations and diagonalizing matrices,* $\omega_j(\boldsymbol{k})$ *and* D.
- $e^{\mathcal{H}_b t}$, *a propagator that accurately solves the interior problem.*
- $k_{\max}$, *the maximal frequency of the problem.*

Algorithm:

`let` $\boldsymbol{u}_d(x, 0)$:= $\boldsymbol{u}_0(x)$ # *Initial condition*

`let` $T_{\text{step}} \leq w/3v_{\max}$ # *Time between filtering operations*

`let` M := $\lceil T_{\max}/T_{\text{step}} \rceil$ #*Number of iterations*

`for` $n \in \{0, 1, \ldots, M = T_{\max}/T_{\text{step}}\}$ `do:`

(1) Propagation step:
`let` $\boldsymbol{v}(x)$:= $e^{\mathcal{H}T_{\text{step}}}\boldsymbol{u}_d(x, nT_{\text{step}})$

(2) Filtering step ($\mathcal{O}_j^{\pm}$ are defined in Section 2.1):
`let` $\boldsymbol{u}_d(\boldsymbol{x}, (n+1)T_{\text{step}})$:= $\left[\prod_{j=1}^{N}(1 - \mathcal{O}_j^{+})(1 - \mathcal{O}_k^{-})\right] \boldsymbol{v}(x)$

(3) `output` $\boldsymbol{u}_d(x, (n+1)T_{\text{step}})$.

The algorithm consists of propagating over time intervals T_{step} too short for even the fastest waves to cross the buffer region and reach the boundary (step 1 inside the for-loop). Step 2 inside the loop consists of filtering outgoing waves, which allows the solution to be propagated further (the waves we filtered are the waves which *would have* reached the boundary. All that remains is to construct the filters, $\mathcal{O}_j^{\pm}$.

2.1. *Construction of the boundary filter*

For a fixed boundary region (say the right boundary), if $\partial_1 \omega_j(\boldsymbol{k}) > 0$, then waves with frequency $\boldsymbol{k}$ are outgoing at this boundary. The outgoing region of *phase space* at the right boundary is:

$$\{(\boldsymbol{x}, \boldsymbol{k}) \in \mathbb{R}^N \times \mathbb{R}^N : \boldsymbol{x}_1 > L \text{ and } \partial_1 \omega_j(\boldsymbol{k}) > 0\}. \tag{4}$$

We construct a projection onto this region. The Heisenberg uncertainty principle makes exact projections impossible, but we can come close. Define:

$$\chi_j^{\pm}(x) = (1/2)[\operatorname{erf}(\sigma^{-1}(x - L - w/3)) - \operatorname{erf}(\sigma^{-1}(x - L - 2w/3))] \tag{5}$$

The parameter $\sigma = O(w/\ln(\delta^{-1})^{1/2})$ and must satisfy Eq. (6) to ensure that $\chi_j^{\pm}(x) < \delta$ for $x_j \notin [\pm L, \pm(L+w)$ or $x_k \notin [-L-w, L+w]$. Multiplication by $\chi_j(x)$ smoothly projects onto the buffer on the j'th side of the box; δ is an error tolerance, and $\star$ denotes convolution.

Define $R_{j,l} = \{\boldsymbol{k} \in \mathbb{R}^N : \partial_{k_j}\omega_l(\boldsymbol{k}) > 0\}$ to be the set of frequencies with the l'th branch of the group velocity pointing right and $R_{j,l,\delta} = \{\boldsymbol{k} \in R_k : d(\boldsymbol{k}, R_{j,l}^C) > k_b\}$ to be the same set excluding a "buffer" region around the place where the group velocity turns around. The buffer ensures that frequency spreading caused $\chi_j(x)$ does not cause waves to turn around. Given k_b, we choose $\sigma \geq O({k_b}^{-1}\ln(\delta^{-1})^{1/2})$ to make $\chi_j(x)$ smooth enough to minimize frequency spreading.

We *assume* that the frequency content of $\widehat{\boldsymbol{u}}(\boldsymbol{k}, t)$ is not contained in $R_{j,l} \setminus R_{j,l,\delta}$. The exact constraints on σ, w and δ are:

$$\begin{aligned} &{k_b}^{-1}\left(\ln(\delta^{-1}) + \ln\left(w^2 L^{N-1} 2^{3N} \sigma^{3N} \pi^{-3N/2}\right)\right)^{1/2} \\ &\qquad\qquad \leq \sigma \leq \frac{w}{\sqrt{\ln(\delta^{-1}) + N\ln(2\sigma\pi^{-1/2})}} \end{aligned} \tag{6}$$

This constraint ensures that the phase space filters are accurate.[14]

Define $P_{j,l,\delta}(k) = (2\sigma\pi^{-1/2})^N e^{-k^2\sigma^2} \star_k \operatorname{diag}[1_{R_{j,1,\delta}}(\boldsymbol{k}), \ldots, 1_{R_{j,n,\delta}}(\boldsymbol{k})]$, which is a smooth projection (in the basis of eigenvectors of $\mathcal{H}$) onto wave-vectors propagating rightward. Then the operator $D^\dagger P(\boldsymbol{k}) D$ is the projection in the basis of wave-vectors. Finally, we define the operator:

$$\mathcal{O}_1^+ = \chi_1^+(x) D^\dagger P(\boldsymbol{k}) D \chi_1^+(x) \tag{7}$$

This operator is an approximate projection onto waves with group velocity pointing outward, and localized in the boundary region.

2.2. *Implementation details*

One useful property of the TDPSF algorithm is that it is compatible with any reasonable interior solver including Fourier spectral methods. Fourier-based spectral methods are highly accurate independent of the timestep; the only errors are frequency aliasing, machine errors and boundary errors[10,13] (which tend to dominate the others). This is the method we use.

The outgoing wave filters are calculated by truncating to the boundary region after multiplying by $\chi_j^{\pm}(\boldsymbol{x})$ and computing an FFT to apply the frequency domain operators[†]. Therefore, as a practical matter, it is useful to take $w = 2^m \delta x$, with δx the lattice spacing in $\boldsymbol{x}$ and $m \in \mathbb{N}$.

The simulations were implemented in Python using numpy and matplotlib [‡]. Source code is available from the webpage of the second author.

3. Examples

3.1. *A warm-up: 1-dimensional Schrödinger equation*

Phase space filters were originally developed for the Schrödinger equation ($u(x,t)$ is a scalar field, and $\mathcal{H} = i\Delta$). In this case there is no diagonalizing operator and $\nabla_k \omega(\boldsymbol{k}) = k$. Outgoing waves at the right boundary are waves of positive frequency, so $P(k) = \left(2\sigma\pi^{-1/2}\right) e^{-k^2/\sigma^2} \star 1_{k>k_b}(k)$ and:

$$\mathcal{O}_1^+ = \chi_1(x)(2\sigma\pi^{-1/2})[e^{-k^2/\sigma^2} \star 1_{k>k_b}(k)]\chi_1(x)$$

We solved the Schrödinger equation on a lattice of 1024 pts ($\delta x = 0.1$) with initial condition $u(x, t = 0) = e^{ikx}e^{-x^2/2\cdot 7^2}$. We measured the errors for various values of δ, taking $\sigma = 1$. The results are plotted in Fig. 1; the error floor at 10^{-8} is due to machine errors. The error tolerance is achieved, except for waves close to $k = 0$ (where the group velocity turns around).

3.2. *Hyperbolic Systems: the Euler and Maxwell equations*

We now consider two cases where the standard PML is unstable. The Euler equations (linearized about a jet flow with mach number M) can be written in the form of Eq. (1) with $\boldsymbol{u} = (p, v_1, v_2)$ (p is pressure, $\boldsymbol{v}$ is fluid velocity):

$$\mathcal{H} = \begin{bmatrix} M\partial_{x_1} & -\partial_{x_1} & -\partial_{x_2} \\ -\partial_{x_1} & M\partial_{x_1} & 0 \\ -\partial_{x_2} & 0 & M\partial_{x_1} \end{bmatrix} \tag{8}$$

Maxwell's equations can be written similarly; with μ a scalar and assum-

---

[†]This causes negligible error since $\chi_j^{\pm}(\boldsymbol{x}) \leq \delta$ for $\boldsymbol{x}$ outside the boundary region.

[‡]Numpy is available from http://numpy.scipy.org, and matplotlib is available from http://matplotlib.sourceforge.net/

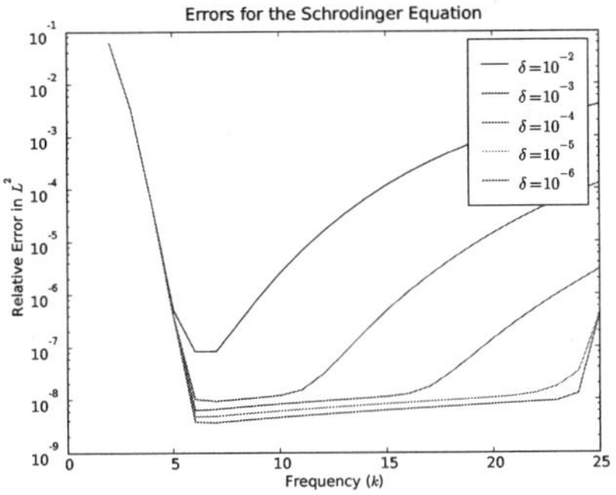

Fig. 1. The relative error for the 1-dimensional test of the Schrodinger equation, as a function of k and the parameter δ.

ing z-independence, they take the form of Eq. (1) with $\boldsymbol{u} = (\sqrt{\mu}\boldsymbol{H}, \sqrt{\epsilon}\boldsymbol{E})^T$

$$\mathcal{H} = \begin{bmatrix} 0 & -\mu^{-1/2}\nabla \times \epsilon^{-1/2} \\ \epsilon^{-1/2}\nabla \times \mu^{-1/2} & 0 \end{bmatrix}, \quad \epsilon = \begin{bmatrix} 1 & b & 0 \\ b & 1 & 0 \\ 0 & 0 & c \end{bmatrix}. \tag{9}$$

We solved these examples on the computational region $[-32, 32]^2$ with $\delta x = 0.125$ (512^2 lattice points), $T_{\max} = 50$ and $k_{\max} = \pi/\delta x = 25.1$. The filter parameters are $w = 16$ (128 lattice points), $T_{\text{step}} = 1.5$ and $\sigma = 1.0$. The initial condition was $u_1(x, t = 0) = r^2 e^{-r^2/9}\cos(Kr)$ with $r = \sqrt{(x-8)^2 + y^2}$ with K varying from 1 to 20, which is localized in frequency near near $|\boldsymbol{k}| = K$. For $K > 4$, L^2 error of 10^{-3} is achieved.

3.3. *Linearized quasi-geostrophic equations*

We now consider the linearized Quasi-Geostrophic equations (also called the midlatitude planetary equations). By linearizing about a streamfunction $\psi(x, t) = -Vy$ (meaning the velocity $\boldsymbol{v} = [-\partial_y\psi, \partial_x\psi]$), the quasi-geostrophic equations take the form:[15]

$$\partial_t\psi - V\partial_x\psi + \tilde{\beta}(-\Delta + F)^{-1}\partial_x\psi = 0$$

This has a dispersion relation $\omega(\boldsymbol{k}) = k_1(V - \tilde{\beta}(|\boldsymbol{k}|^2 + F)^{-1})$. Here, V is the mean wind, F is a constant proportional to f^2/g, with f the rotation frequency of the earth, g the gravitational attraction and $\tilde{\beta} = FV + \beta$ and $\beta = R\cos(\varphi)$ where R is the radius of the earth and φ is the latitude[§]. The

[§]We work in the β-plane approximation, i.e., we study the equations in spherical coordinates, expand trigonometric terms (functions of latitude and longitude) in a Taylor series and work on a small cartesian region.

numerical parameters are the same as above, and we took $V = 1$, $F = 10$, $\beta = 100$; the results are displayed in Fig. 3. Although the results are not as good as for the Euler or Maxwell equations, they are acceptable, and can be improved by taking a larger buffer region. This is due primarily to the non-locality of the equation, not the anisotropy.

In,[7] it is proven for hyperbolic systems that, the PML on the i-th side is stable only when $k_i v_{g,i}(\boldsymbol{k}) > 0$. While the anisotropic Euler and Maxwell equations do not satisfy this criteria, the PML can be made stable for those cases (essentially by completing the square). The quasi-geostrophic equations can not be fixed in this way.

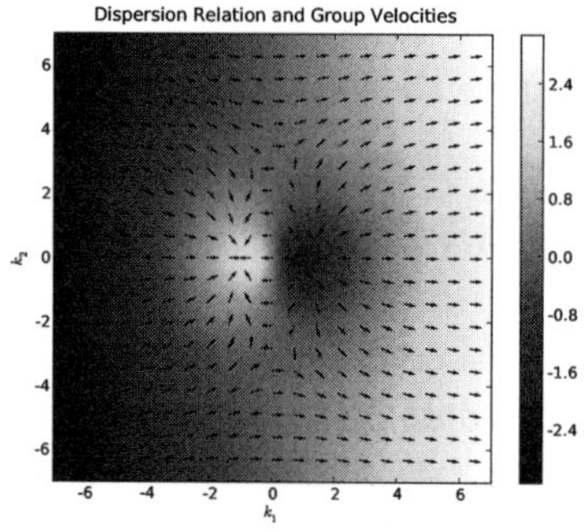

Fig. 2. The dispersion relation and group velocities for the quasigeostrophic equation.

4. Stability

We have the following rigorous estimate[14] regarding Algorithm 1's stability:

$$\|\boldsymbol{u}_d(x,t)\|_{L^2} \leq \|\boldsymbol{u}(x,t')\|_{L^2} \text{ if } t > t' \tag{10}$$

To test this, we solved the Euler, Maxwell and Schrödinger equations up to time $t = 2000$ and measured the energy,[14] confirming the inequality (10).

5. The Low Frequency Problem

Figures 1 and 3 indicate that the TDPSF performs poorly for waves with low frequency. Increasing the width of the filter imposes a computational cost of order $O(k_{\max}/k_b)$ just to resolve the buffer, which is undesirable.

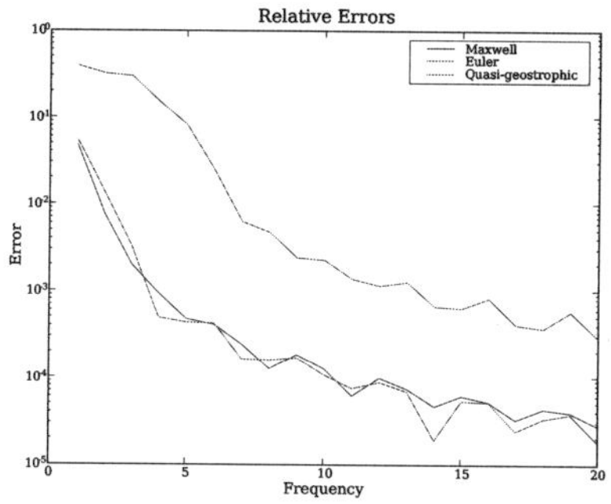

Fig. 3. The relative errors (measured in various norms) for Euler and Maxwell systems as a function of the frequency of the initial condition.

This can be remedied by a multiscale method which imposes cost only $O(\log_2(k_{\max}/k_b))$; we believe this is close to the best possible.[13]

Acknowledgments

We thank Peter Petropoulos for pointing out to us the problem of PML instability and Tom Hagstrom for comments which led to Section 3.3.

References

1. G. Alpert, L. Greengard and T. Hagstrom, *J. Comput. Phys.* **180**, 270 (2002).
2. G. Bayliss and E. Turkel, *Comm. Pure Appl. Math.* **33**, 707 (1980).
3. B. Engquist and A. Majda, *Proc. Nat. Acad. Science U.S.A* **77**, 1765 (1977).
4. S. Jiang and L. Greengard, *Comm. Pure Appl. Math.* **61**, 261 (2008).
5. J.-P. Berenger, *J. Comput. Phys.* **114**, 185 (1994).
6. F. Hu, *J. Comput. Phys.* **129**, 201 (1996).
7. E. Becache, S. Fauqueux and P. Joly, *J. Comput. Phys.* **188**, 399 (2003).
8. D. G. S. Abarbanel and J. Hesthaven, *Journal of Scientific Computing* **17**, (2002).
9. P. G. Becache, E. Petropoulos and S. D. Gedney, *IEEE Trans. Ant. Prop.* **52**, (2004).
10. A. Soffer and C. Stucchio (2006), in preparation.
11. A. Soffer and C. Stucchio, *J. Comput. Phys.* **225**, 1218(2007).
12. C. Stucchio, Selected Problems in Quantum Mechanics, PhD thesis, Rutgers University, (Piscataway, NJ, 2008).
13. A. Soffer and C. Stucchio, *Comm. Pure and Appl. Math.* (Accepted).
14. A. Soffer and C. Stucchio, (Submitted) (2008).
15. A. Majda, *Introduction to PDES and Waves for the Atmosphere and Ocean Waves* (American Mathematical Society, Providence, RI, 2003)

ASYMPTOTIC APPROXIMATIONS IN FINANCIAL MATHEMATICS

RICHARD JORDAN

Department of Mathematics, Statistics and Computer Science
University of Illinois at Chicago, Chicago, IL 60607, USA
E-mail: jordan@math.uic.edu

CHARLES TIER

Department of Applied Mathematics
Illinois Institute of Technology Chicago, IL 60616, USA
E-mail: ctier@itt.edu

There has been growing interest in more sophisticated models for pricing financial derivatives that go beyond the Black-Scholes-Merton Model. We consider models in which the volatility of the underlying financial asset is no longer constant but may be a deterministic or stochastic function. Exact and asymptotic methods such as the ray method are used to find simple, useful pricing formulas.

Keywords: financial derivatives, CEV model, asymptotic methods

1. Introduction

The rapid growth of financial derivatives during the last thirty years has led practitioners to develop more detailed and realistic models of market behavior. The pricing and hedging of derivatives has its origins in the Nobel prize winning work of Black, Scholes, and Merton (BSM).[1,2] The BSM theory assumes the underlying asset $S(t)$ follows geometric Brownian motion. If $c(S,t)$ is the price of a European call option at time t for an asset with price $S(t)$, strike price K, and at expiry T with a payoff of $\max(S-K,0)$ then BSM theory leads to the celebrated formula

$$c(S,t;\sigma,K) = SN(d_1) - e^{-r(T-t)}KN(d_2) \tag{1}$$

where $N(x)$ is the cumulative normal distribution function and

$$d_1 = (\ln(S/K) + (r + \frac{1}{2}\sigma^2)(T-t))/\sigma\sqrt{T-t}, \quad d_2 = d_1 - \sigma\sqrt{T-t}.$$

Here r is the risk-free interest rate and σ is the constant volatility. The formula (1) provides not only the price of the call option but also a scheme for replicating the option in terms of the underlying asset and bonds. The parameters in Eq. (1) are all easily observable except for the volatility σ which must be estimated in some fashion.

A wonderful aspect of market-traded options is that the prices are listed for the public. Let $c_{market}(K,T)$ be the observed market price of a call option with expiry T and strike price K. We can use these prices to determine what the market implies for the value of the volatility $\sigma_{implied}(K,T)$ by solving the inverse problem

$$c(S,t;\sigma_{implied},K) = c_{market}, \tag{2}$$

generating an *implied volatility surface.* BSM theory predicts that the volatility surface is constant. However, since the crash of 1987 and the LTCM meltdown (1988), the market volatility has increased dramatically and the implied volatility surface is not constant but instead has a *smile* and/or *skew.* The implications are that traders need to be able to better hedge volatility risk and the BSM theory is deficient.

2. Volatility Index

In 2003, the Chicago Board Options Exchange established a new volatility index (VIX) for traders and the public to use as a measure of market volatility (fear index). It is based on the weighted average of option prices across all strikes at two nearby maturities. This strip of options replicates the realized variance of the S & P 500 index as follows. Let $\{S_0, S_1, \ldots, S_N\}$ be a sequence of observations of the value of the S & P 500 index with the realized variance defined in terms of the continuous return by

$$v_R^2 = \frac{c^2}{N}\sum_{i=1}^{N}(\ln\frac{S_i}{S_{i-1}})^2$$

where c^2 is an annual conversion factor. The VIX is defined to be the next 30 day calendar value using the formula

$$VIX^2 = \mathrm{E}_0^Q[v_R^2]. \tag{3}$$

The realized variance is then replaced by a continuous approximation

$$v_R^2 \approx \sigma_R^2 = \frac{1}{T}\int_0^T \sigma^2 dt$$

and assuming the asset price follows a stochastic differential equation with Brownian motion then

$$\mathrm{E}_0^Q[v_R^2] \approx \mathrm{E}_0^Q[\sigma_R^2] = \frac{2}{T}\left[\mathrm{E}_0^Q[\int_0^T \frac{dS}{S}] - \mathrm{E}_0^Q[\ln\frac{S(T)}{S(0)}]\right] \tag{4}$$

The second term is the payoff of the log contract[3] which is known to be related to a strip of options[4] with maturity T so that

$$\mathrm{E}_0^Q[\sigma_R^2] = \frac{2}{T}e^{rT}\left(\int_0^{S_0}\frac{1}{K^2}p(S,t;K)dK + \int_{S_0}^{\infty}\frac{1}{K^2}c(S,t;K)dK\right)$$

This is the basis of the VIX when $T = 1/12$.

3. Variance Swap

A variance swap[5] is a forward contract in which holder exchanges realized variance v_R^2 for a variance strike K_{var}. Thus the payoff is $(v_R^2 - K_{var})$ so that the value of the variance swap is

$$V(t) = e^{-r(T-t)}\left(\mathrm{E}_t^Q[v_R^2] - K_{var}\right).$$

When the contract is initiated ($t = 0$) it has zero value so that the variance rate (future realized variance) is determined using Eq. (4) with $V(0) = 0$ as

$$K_{var} = \mathrm{E}_0^Q[v_R^2] \approx \mathrm{E}_0^Q[\sigma_R^2] = \frac{2}{T}\left[\mathrm{E}_0^{\mathbf{Q}}\left[\int_0^T\frac{dS}{S}\right] - e^{rT}L(S,0)\right]. \tag{5}$$

Here $L(S,0) = e^{-rT}\mathrm{E}_0^Q[\ln\frac{S(T)}{S(0)}]$ is the value of the log contract at $t = 0$. To price a variance swap, we must find $K_{var} \approx \mathrm{E}_0^Q[\sigma_R^2]$ which we will do using the log contract directly without using a strip of options that may not be available.

Over the life of the contract, the variance consists of two parts: the realized variance from the start of the contract till the present time and the unrealized portion up till expiry, i.e. $\mathrm{E}_t^Q[v_R^2]$.

4. Log Contract and CEV Process

We will investigate the log contract when the underlying asset follows the Constant Elasticity of Variance (CEV) process[6]

$$dS = rSdt + \delta S^{\beta+1}dW, \qquad S > 0, \quad t \in [0,\infty).$$

Here the volatility is deterministic and no longer constant as in the BSM model, which is the special case $\beta = 0$. The equation for the price $L(S,t)$ of the log contract is

$$\frac{\partial L}{\partial t} + \frac{1}{2}\delta^2 S^{2\beta+2}\frac{\partial^2 L}{\partial S^2} + rS\frac{\partial L}{\partial S} - rL = 0,\ L(S,T) = \ln\left(\frac{S}{S_0}\right),\ S_0 = S(0). \tag{6}$$

By applying the following sequence of transformations[7] to Eq. (6)

$$x = \frac{(Se^{r(T-t)})^{-2\beta}}{\delta^2\beta^2};\quad L = e^{-r(T-t)}u;$$
$$\tau = \frac{1}{2r\beta}\left[1 - e^{-2r\beta(T-t)}\right],\quad t = T \to \tau = 0,$$

we obtain

$$\frac{\partial u}{\partial \tau} - 2x\frac{\partial^2 u}{\partial x^2} - \left(2 + \frac{1}{\beta}\right)\frac{\partial u}{\partial x} = 0 \tag{7}$$

with the initial condition

$$u(x,0) = -\frac{1}{2\beta}\ln\left(x\delta^2\beta^2\right) - \ln(S_0).$$

The underlying stochastic process is the squared Bessel process[8]

$$dx = \left(2 + \frac{1}{\beta}\right)d\tau + 2\sqrt{x}dW.$$

The PDE is singular at both $x = 0$ and $x = \infty$ and so boundary conditions may not be arbitrary.

The Green's function[8] $p(x, \hat{x}, \tau)$ for Eq. (7) is

$$p(x,\hat{x},\tau) = \frac{1}{2\tau}\left(\frac{\hat{x}}{x}\right)^{1/(4\beta)}\exp\left(-\frac{\hat{x}+x}{2\tau}\right)I_{1/(2|\beta|)}\left(\frac{\sqrt{\hat{x}x}}{\tau}\right),\quad \tau \geq 0$$

where $x = x(\tau)$, $\hat{x} = x(0)$ and $I_\nu(\cdot)$ is the modified Bessel function. Recall that $\tau = 0$ corresponds to the expiry $t = T$.

4.1. *Pricing Formulas*

The exact integral representation for the price of the log contract is

$$L(S,t) = e^{-r(T-t)}\left[\int_0^\infty \ln(\frac{\hat{S}}{S_0})p(\hat{S},S,t)d\hat{S}\right],\quad \beta > 0$$
$$= e^{-r(T-t)}\left\{-\frac{1}{2\beta}[I(z) + \ln(2\tau\delta^2\beta^2)] - \ln(S_0)\right\}$$

where

$$I(z) = \int_0^\infty \ln(\hat{z}) \left(\frac{\hat{z}}{z}\right)^{1/(4\beta)} e^{-\hat{z}-z} I_{1/(2|\beta|)}\left(2\sqrt{\hat{z}z}\right) d\hat{z},$$

with

$$z = \frac{r\left[Se^{r(T-t)}\right]^{-2\beta}}{\beta\delta^2\left[1 - e^{-2r\beta(T-t)}\right]}, \quad r \neq 0.$$

To find $L(S,t)$, we must evaluate $I(z)$. For special cases, we are able to find a simple analytic formula. Otherwise, we construct approximate or asymptotic results. One might try integrating $I(z)$ numerically but this proves difficult for large z or $t \approx T$, i.e. near expiration.

4.2. $\beta > 0$

We are able to find the following expansion[9] for small z

$$I(z) = e^{-z} \sum_{k=0}^{\infty} \frac{z^k}{k!} \Psi\left(\frac{1}{2\beta} + 1 + k\right)$$

where $\Psi(x)$ is the Psi function. We sum the series for $I'(z)$ and obtain the integral representation

$$I'(z) = \frac{e^{-z}}{z^{1/(2\beta)+1}} \int_0^z t^{1/(2\beta)} e^t \, dt.$$

If $1/(2\beta) + 1 = m$ then

$$I'(z) = \frac{e^{-z}}{z^m} \int_0^z t^{m-1} e^t \, dt$$

which we integrate by parts and then integrate to find $I(z)$ leading to an exact expansion useful for large z

$$I(z) = \ln(z) + \sum_{k=1}^{m-1} \frac{(-1)^{k+1}(m-1)!}{k(m-1-k)!} \frac{1}{z^k}.$$

Using Eq. (5) and that $\mathrm{E}_0^{\mathbf{Q}}\left[\int_0^T \frac{dS}{S}\right] = rT$ we find that

$$K_{var} = \frac{1}{\beta T} \sum_{k=1}^{m-1} \frac{(-1)^{k+1}(m-1)!}{k(m-1-k)!} \left(\frac{\beta\delta^2}{r}\left[1 - e^{-2r\beta T}\right]\left[Se^{rT}\right]^{2\beta}\right)^k.$$

If $\beta > 0$ and $1/(2\beta) + 1 \neq m$, we obtain an asymptotic expansion

$$I(z) \sim \ln(z) + \sum_{k=1}^{\infty} \frac{(-1)^{k+1}}{kz^k} \frac{1}{2\beta}\left(\frac{1}{2\beta} - 1\right) \cdots \left(\frac{1}{2\beta} - k + 1\right), \quad z \to \infty$$

and hence

$$K_{var} \sim \frac{1}{\beta T}\sum_{k=1}^{\infty}\frac{(-1)^{k+1}}{k}\frac{1}{2\beta}\left(\frac{1}{2\beta}-1\right)\cdots\left(\frac{1}{2\beta}-k+1\right)$$
$$\left(\frac{\beta\delta^2}{r}\left[1-e^{-2r\beta T}\right]\left[Se^{rT}\right]^{2\beta}\right)^k.$$

4.3. $\beta < 0$

This case presents difficulties since $S = 0$ is an absorbing boundary and default of the asset is possible. The probability of default can be computed but the log payoff is not defined at $S = 0$. In Ref. 9, we present an analysis assuming a finite value of the log contract if default occurs. We also present an analysis of a *shifted* CEV model[8] in which default occurs at $S_{\min} > 0$. This avoids the problem of the log contract not being defined at $S = 0$.

5. Stochastic CEV Model

Another model that has been successful in accounting for the observed volatility smile/skew is the stochastic CEV or SABR model.[10] In this model, the forward price of the asset $F(t)$ as well as the volatility σ are stochastic and follow

$$\begin{aligned} dF &= \sigma F^{\beta+1} d\widetilde{W}_1 \\ d\sigma &= \nu\sigma dW_2 \end{aligned}$$

where $\widetilde{W}_1$ and W_2 are correlated Brownian motions with $\mathrm{E}[d\widetilde{W}_1 dW_2] = \rho\, dt$ and $\rho \in [-1, 1]$. An equivalent set of equations is

$$\begin{aligned} dF &= \sqrt{1-\rho^2}\sigma F^{\beta+1} dW_1 + \rho\sigma F^{\beta+1} dW_2 \\ d\sigma &= \nu\sigma dW_2 \end{aligned}$$

where W_1 and W_2 are now independent Brownian motions.

We are interested in risk-neutral pricing of derivatives such as the log contract in which case the Price $= e^{-r(T-t)}\mathrm{E}_t^Q[\text{payoff}]$. To compute the expectation, we first must construct the density function $p = p(\widehat{F}, \widehat{\sigma}, T; F, \sigma, t)$ for the stochastic system above. Here the initial values are $F = F(t)$ and $\sigma = \sigma(t)$ while the final values are $\widehat{F} = F(T)$, $\widehat{\sigma} = \sigma(T)$. The density function satisfies the forward equation

$$\frac{\partial p}{\partial T} = \frac{1}{2}\frac{\partial^2}{\partial \widehat{F}^2}\left[\widehat{\sigma}^2\widehat{F}^{2\beta+2}p\right] + \frac{1}{2}\frac{\partial^2}{\partial\widehat{\sigma}^2}\left[\nu^2\widehat{\sigma}^2 p\right] + \frac{\partial^2}{\partial\widehat{F}\partial\widehat{\sigma}}\left[\rho\nu\widehat{\sigma}^2\widehat{F}^{\beta+1}p\right] \tag{8}$$

with boundary conditions[11] and the initial condition

$$\lim_{T\to t} p(\widehat{F}, \widehat{\sigma}, T; F, \sigma, t) = \delta(\widehat{F} - F)\delta(\widehat{\sigma} - \sigma).$$

The domain is $0 < \widehat{F} < \infty$ and $0 < \widehat{\sigma} < \infty$. Our goal is to construct asymptotic solutions for p.

5.1. *Special Case: $\rho = 0$, $\beta = -1$*

We first look at a special case of Eq. (8) with $\rho = 0$ and $\beta = -1$ for which an exact solution is known.[12] We introduce the new variables $\tau = \nu^2(T - t)/2$, $\widehat{F} \to \widehat{y}$, $\widehat{x} = \widehat{\sigma}/\nu$ into Eq. (8) to obtain

$$\frac{\partial p}{\partial \tau} = \widehat{x}^2 \frac{\partial^2 p}{\partial \widehat{y}^2} + \frac{\partial^2 (\widehat{x}^2 p)}{\partial \widehat{x}^2}, \quad 0 < \widehat{x} < \infty, \quad -\infty < \widehat{y} < \infty$$

with initial condition

$$\lim_{\tau\to 0} p(\widehat{y}, \widehat{x}, \tau; y, x, 0) = \delta(\widehat{y} - y)\delta(\widehat{x} - x).$$

This corresponds to diffusion on a surface of negative curvature[12,13] - Poincaré Plane $\mathbb{H}^2$.

An exact formula for the probability density function $p(\widehat{y}, \widehat{x}, \tau; y, x)$ was found by McKean[12] as

$$p(\widehat{y}, \widehat{x}, \tau; y, x) = \frac{e^{-\tau/4}\sqrt{2}}{(4\pi\tau)^{3/2}\widehat{x}^2} \int_{\varphi}^{\infty} \frac{z e^{-z^2/4\tau}}{\sqrt{\cosh(z) - \cosh(\varphi)}} dz$$

where

$$\varphi = \cosh^{-1}\left(1 + \frac{(x - \widehat{x})^2 + (y - \widehat{y})^2}{2\widehat{x}x}\right).$$

Here φ is the distance between points (x, y) and $(\widehat{x}, \widehat{y})$ on Poincaré plane.

5.2. *Ray Solution for short time*

For general values of ρ and β no exact solution to Eq. (8) is available. Instead we will construct an asymptotic approximation in the form the ray solution for short time.[14–16] We seek a solution of Eq. (8) for $\tau \ll 1$ in the form

$$p(\widehat{F}, \widehat{\sigma}, \tau; F, \sigma, 0) \sim e^{-\widetilde{\varphi}^2/4\tau} \sum_{n=0}^{\infty} \tau^{n-1} Z_n(\widehat{F}, \widehat{\sigma}; F, \sigma).$$

This leads to an eikonal equation

$$\widehat{x}^2 \left[\varphi_{\widehat{y}}^2 + \varphi_{\widehat{x}}^2 - 2\rho\varphi_{\widehat{x}}\varphi_{\widehat{y}}\right] = 1$$

with $\widehat{x} = \widehat{\sigma}/\nu$, $\widehat{y} = \widehat{F}^{-\beta}/\beta$, $\varphi = \nu\widehat{\varphi}/\sqrt{2}$. If $\rho = 0$ this corresponds to the eikonal equation for McKean's problem. There is a transport equation for Z_0.[11] The solution of eikonal equation is found by the method of characteristics to be

$$\varphi = \cosh^{-1}\left(1 + \frac{(x-\widehat{x})^2 + 2\rho(x-\widehat{x})(y-\widehat{y}) + (y-\widehat{y})^2}{2(1-\rho^2)\widehat{x}x}\right).$$

The solution to the transport equation is given in Ref. 11. Unfortunately, the ray solution is not valid near $\widehat{F} = 0$ and a boundary layer analysis is needed to satisfy the appropriate boundary condition at $\widehat{F} = 0$.[11]

References

1. F. Black and M. Scholes, *Journal of Political Economy* **81**, 637 (1973).
2. R. Merton, *Bell Journal of Economics and Mamagement Science* **4**, 141 (1973).
3. A. Neuberger, *Journal of Portfolio Management* **Winter 1994**, 74 (1994).
4. P. Carr and D. Madan, *Volatility: New Estimation Techniques for Pricing Derivatives* (Risk Publications, 1998), ch. 29, Towards a Theory of Volatility Trading.
5. K. Demeterfi, E. Derman, M. Kamal and J. Zou, *Journal of Derivatives* **Summer**, 9 (1999).
6. J. Cox and S. Ross, *Journal of Financial Economics* **3**, 145 (1975).
7. D. Davydov and V. Linetsky, *Management Science* **47**, 949 (2001).
8. C. Albanese and G. Campolieti, *Advanced Derivatives Pricing and Risk Management*, 1st edn. (Elsevier Academic Press, 2006).
9. R. Jordan and C. Tier, The variance swap contract under the cev process(August, 2007).
10. P. Hagan, D. Kumar, A. Lesniewski and D. Woodward, *Wilmott Magazine* **Sept**, 84 (2002).
11. R. Jordan, Asymptotic methods applied to finance: Equity and volatility derivatives, PhD thesis, University of Illinois at Chicago, (Dept. of Mathematics, 2008).
12. H. McKean, *Journal of Differential Geometry* **4**, 359 (1970).
13. P. Hagan, , A. Lesniewski and D. Woodward, Probability distribution in the sabr model of stochastic volatility, Working Paper(June, 2004).
14. J. Cohen and R. Lewis, *Journal of the Institute of Mathematics and Its Applications* **3**, 266 (1967).
15. C. Tier and J. Keller, *SIAM Journal of Applied Mathematics* **34**, 549 (1978).
16. P. Henry-Labordère, A general asymptotic implied volatility for stochastic volatility models, Working Paper(May, 2005).

AUTHOR INDEX